ENGINEERING INSTRUCTION

for High-Ability Learners
in K-8 Classrooms

ENGINEERING
INSTRUCTION

for High-Ability Learners in K-8 Classrooms

Edited by
Debbie Dailey, Ed.D.,
and Alicia Cotabish, Ed.D.

NATIONAL ASSOCIATION FOR
Gifted Children
SERVICE PUBLICATION

Routledge
Taylor & Francis Group

NEW YORK AND LONDON

First published in 2017 by Prufrock Press Inc.

Published 2021 by Routledge
605 Third Avenue, New York, NY 10017
2 Park Square, Milton Park, Abingdon, Oxon OX14 4RN

Routledge is an imprint of the Taylor & Francis Group, an informa business

Library of Congress Cataloging-in-Publication Data

Names: Dailey, Debbie, 1963- editor. | Cotabish, Alicia, editor.
Title: Engineering instruction for high-ability learners in K-8 classrooms / edited by Debbie Dailey and Alicia Cotabish.
Other titles: Engineering instruction for high-ability learners in K-eight classrooms
Description: New York, NY: Taylor & Francis, [2017] | "A Service Publication of the National Association for Gifted Children." | Includes bibliographical references.
Identifiers: LCCN 2016036075| ISBN 9781618216144 (pbk.) | ISBN 9781618216168 (epub) | ISBN 9781618216151 (pdf)
Subjects: LCSH: Engineering--Study and teaching (Elementary) | Gifted children--Education (Elementary)
Classification: LCC LB1594 .E544 2017 | DDC 372.35/8--dc23
LC record available at https://lccn.loc.gov/2016036075

ISBN 13: 978-1-0321-4403-0 (hbk)
ISBN 13: 978-1-6182-1614-4 (pbk)

DOI: 10.4324/9781003234951

Table of Contents

ACKNOWLEDGEMENTS

Many people have assisted with the efforts in developing this book. They include the expert authors of the chapters, the leadership and staff of the National Association for Gifted Children (NAGC), and the NAGC Publication Committee. This book has also been strengthened through a rigorous review process. We want to thank the reviewers, Amy Sedivy-Benton, Jeff Danielian, Lesley Evans, James Fetterly, Janine Firmender, Adrienne Gifford, Vernard Henley, Nykela Jackson, Suzanne Mitchell, Michael Mills, Karen Rambo-Hernandez, and Bess Wilson, who took time to provide us with valuable feedback and direction. Finally, this project would not be possible without the help of Jane Clarenbach, Director of Public Education at NAGC. Jane was always available to review chapters and give advice and direction. She was invaluable to this project.

On a personal level, we would like to thank our families and our university for allowing us the time and support to complete this project.

FOREWORD

Developing Engineering Innovators Among Our Most Talented Students

Well-documented trends have been reported nationally of declining interest, poor preparedness, lack of diverse representation, and low persistence of U.S. students in STEM (science, technology, engineering, and mathematics) disciplines. Research suggests that the reasons for these trends are not due to students' performance, but are closely linked to an early loss of interest in science, perceptions of poor teaching, selecting a STEM major with insufficient information about the career, and feeling overwhelmed by the pace and load of the curriculum (Ali, Yager, Hacieminoglu, & Caliskan, 2013; Bitan-Friedlander, Drefus, & Milgrom, 2004; Johnson, Kahle, & Fargo, 2006). To reverse these trends and attract students to STEM careers, we must pursue targeted initiatives. First, it is imperative that we increase the level of academic preparation among the pool of students interested in STEM careers so they are prepared for the rigors of STEM education. Second, we need to reform the pedagogy and culture of teaching in K–12 schools in order to create exciting and engaging STEM-focused learning experiences for all students. Finally, we must develop strong mentoring and support programs for students to ensure that those students maintain their interest in STEM disciplines and experience success in these areas throughout their K–12 schooling.

However, teachers often lack adequate support, including appropriate professional development, as well as interesting and intriguing curricula. In addition, school systems lack tools for assessing progress and rewarding success. As a result, too many students conclude early in their education that STEM subjects are boring, too difficult, or unwelcoming, leaving them ill-prepared to meet the challenges that will face their generation, their country, and the world.

Despite these troubling signs, we have great strengths on which we can draw. A growing body of research has illuminated how children learn about STEM, making it possible to devise more effective instructional materials and teaching strategies. The National Research Council and other organizations have summarized this research in a number of influential reports and have made recommendations concerning the teaching of mathematics, science, and engineering. These reports transcend tired debates about conceptual understanding versus factual recall versus procedural fluency. They emphasize that students learning science, mathematics, and engineering need to acquire all of these capabilities, because they support each other. Secondly, a clear bipartisan consensus has emerged from the need for general education reform and the importance of STEM education in particular. These more recent events have created promising opportunities within the STEM disciplines. Furthermore, the recent signing and inclusion of gifted and talented students in key sections of the Every Student Succeeds Act (ESSA) has been a monumental success and gives schools the leeway to devote considerable attention and resources to develop talent among this often ignored group of students.

To reiterate how we identify and develop talent among our most able students, we can turn to the recommendations put forth by the National Science Board (NSB; 2010). Its three keystone recommendations, supported by the best available research, state that we must:

1. **Provide opportunities for excellence.** We cannot assume that our Nation's most talented students will succeed on their own. Instead, we must offer coordinated, proactive, sustained formal and informal interventions to develop their abilities. Students should learn at a pace, depth, and breadth commensurate with their talents and interests and in a fashion that elicits engagement, intellectual curiosity, and creative problem solving—essential skills for future innovation.

2. **Cast a wide net** to identify all types of talents and to nurture potential in all demographics of students. To this end, we must develop and implement appropriate talent assessments at multiple grade levels and prepare educators to recognize potential, particularly among those indi-

viduals who have not been given adequate opportunities to transform their potential into academic achievement.

3. **Foster a supportive ecosystem** that nurtures and celebrates excellence and innovative thinking. Parents/guardians, education professionals, peers, and students themselves must work together to create a culture that expects excellence, encourages creativity, and rewards the successes of all students regardless of their race/ethnicity, gender, socioeconomic status, or geographical locale. (pp. 2–3; emphasis in original)

To this end, I am very excited about this book and what it brings to the table with regard to identifying and developing engineering talent in K–8 classrooms and how it supports the NSB recommendations. As the editor of the Exemplary Science Programs (ESP) monograph series for the National Science Teachers Association, I have been actively involved in promoting STEM, particularly science teaching, and program and talent development at all levels along the K–20 education spectrum. I have dedicated my career to informing the public and trying to improve the discipline. This book fills a gap and certainly moves the field along by addressing the "E" in STEM through creating learning experiences for our most promising and able students. Its 13 chapters include an array of engineering education strategies and suggestions from notable educators in engineering and gifted education. These include supports for developing spatial ability, designing engineering experiences, integrating technology, creating professional development and other learning experiences for teachers, developing and integrating engineering curriculum, and assessing engineering activities, among others.

In summary, when teachers change their teaching, which includes increasing discipline-specific knowledge, skills, and pedagogical practices, student interest increases—and more students aspire to STEM-related careers, including engineering. Inspired teaching must occur, and student learning experiences must be challenging, engaging, and include advanced problem solving. Of course, professional development supports from schools and their administrators are required for these changes to take place. Lastly, educators and their support personnel must plan the work—and work the plan!

Robert E. Yager, Ph.D.
Professor Emeritus, University of Iowa

REFERENCES

Ali, M. M., Yager, R. E., Hacieminoglu, E., & Caliskan, İ. (2013). Changes in student attitudes regarding science when taught by teachers without experiences with a model professional development program. *School Science and Mathematics, 113,* 109–119.

Bitan-Friedlander, N., Drefus, A., & Milgrom, Z. (2004). Types of "teachers in training": The reactions of primary school science teachers when confronted with the task of implementing an innovation. *Teaching and Teacher Education, 20,* 607–619. doi:10.1016/j.tate.2004.06.007

Johnson, C. J., Kahle, J. B., & Fargo, J. D. (2006). Effective teaching results in increased science achievement for all students. *Science Education, 91,* 371–383. doi:10:1002/sce.20195

National Science Board. (2010, May 5). *Preparing the next generation of STEM innovators: Identifying and developing our nation's human capital.* Retrieved from http://www.nsf.gov/nsb/publications/2010/nsb1033.pdf

INTRODUCTION

DOI: 10.4324/9781003234951-1

1

Designing Innovative Engineering Instruction for High-Ability Learners in K–8 Classrooms

Debbie Dailey and Alicia Cotabish

In 2013, Rothwell reported that 20% of all U.S. jobs required a significant background in science, technology, engineering, or mathematics (STEM). Even more astonishing, 26 million jobs in the U.S. are dependent upon employees with some background in STEM. Due to these statistics and reports that our educational system is not adequately preparing workers for STEM jobs (Change the Equation, 2012), the Next Generation Science Standards (NGSS) were developed, released, and adopted by many states in hopes to improve STEM education (NGSS Lead States, 2013). The authors of *A Framework for K–12 Science Education*, the foundation for the NGSS, recommended the standards focus on disciplinary core ideas, crosscutting or overarching concepts, and science and engineering practices (National Research Council [NRC], 2012). This allows students to have an integrated study of science through content, concepts, and processes, a strategy long recommended by VanTassel-Baska (1992, 1998).

With the recent and continuous technological advancements in society, NGSS sought to add engineering practices to the standards (NGSS Lead States, 2013). The goal is not necessarily to increase the number of engineers but instead to use science and engineering practices to teach science content so that students will gain "sufficient knowledge of the practices, crosscutting concepts, and core ideas of science and engineering to engage in public discussions on science-related issues, to be critical consumers of scientific information related to their everyday lives, and to continue to learn about science throughout their lives" (NRC, 2012, p. 12). Additionally, using engineering practices to teach science content engages students in active learning, allowing them to investigate the natural world and develop solutions to meaningful problems that they may encounter in the real world.

There are eight science and engineering practices (Table 0.1) in the NGSS that portray the real-world actions of scientists and engineers as they investigate the natural world or design and build models and systems to address a need (Achieve, Inc., 2014). The practices are designed to be integrated with content (disciplinary core ideas) and crosscutting concepts to explain a particular phenomenon and should not be taught in isolation. For example, the practices should not be presented in Chapter 1 of the textbook and never revisited again—as often occurred in past years with the scientific method. Instead, practices must be embedded in content to mimic real-world science and make the content more relevant and engaging to students. Teachers of advanced and gifted and talented students are not unfamiliar with engineering design practices—as many gifted students have engaged in activities such as building bridges or designing egg-drop containers. However, many times the activities were just activities and not embedded in the content of science or any other discipline. Throughout this book, we want to emphasize the importance of engaging students in the practices of science and engineering while addressing the content standards.

PURPOSE OF THE BOOK

The purpose of this book is to assist educators and practitioners in addressing engineering standards in their general and gifted K–8 classrooms, with a particular focus on students with a high affinity, interest, and/or talent for engineering design. As former classroom and gifted teachers and now teacher educators and researchers, we hope that practitioners will be able to use this book to overcome barriers they may face in implementing engineering in their classrooms. We are excited to bring together notable authors and researchers to assist us in this

TABLE 0.1
Science and Engineering Practices

Asking Questions (for science) and Defining Problems (for engineering)
Developing and Using Models
Planning and Carrying Out Investigations
Analyzing and Interpreting Data
Using Mathematics and Computational Thinking
Constructing Explanations (for science) and Designing Solutions (for engineering)
Engaging in Argument From Evidence
Obtaining, Evaluating, and Communicating Information

From National Research Council (2012).

endeavor—to help teachers engage students in engineering. As a previous director and coordinator of a United States Department of Education grant, STEM Starters, we have seen firsthand how difficult it is for teachers, in particular elementary teachers, to add science and engineering into their daily curriculum. There are many barriers that teachers face when seeking to improve science programs in their classrooms, including "(a) time constraints and scheduling conflicts, (b) insufficient resources, (c) inadequate teacher science knowledge and skills, and (d) poor teacher confidence" (Dailey, 2015, p. 21). With this in mind, the authors devote 13 chapters to focusing on applications of engineering design and practices, with special attention paid to designing engineering curriculum and instruction, integrating technology into engineering instruction, and assessing engineering practices for high-ability learners.

CHAPTER AUTHORS

This book has a wonderful mix of early career and distinguished senior scholars. Many of the authors come from gifted education, but science and engineering content experts are also represented. Additionally, a highly renowned science education scholar, Dr. Robert Yager, provided the forward for this book. We are honored to work with each of them.

BOOK ORGANIZATION

The book is organized into four sections. Section 1 addresses key components of engineering instruction for K–8 high-ability learners. In this section, Dr. Steve Coxon discusses the role of spatial ability in engineering, and Drs. Eric and Rachel Mann examine engineering design using gifted pedagogy. Dr. Laurie Croft suggests ways to integrate inventiveness, innovation, and creativity into engineering, and Dr. Rachelle Miller and Ms. Callie Slider encourage educators to engage students in art while teaching engineering and science.

Section 2 explicates how to use cutting-edge technology in engineering curriculum and instruction. Dr. Jason Trumble discusses ways to incorporate innovative technology tools, such as 3-D printers, and Ms. Irene Lee and Dr. April Degannaro suggest how to use computer science and coding for project-based engineering instruction. Additionally, Ms. Krissy Venosdale and Dr. Brian Housand present how to use the ever-popular Maker Movement to facilitate engineering problem solving in creative and gifted learners.

Section 3 focuses on the curriculum and how to design or integrate engineering practices into classroom instruction. Ms. Michelle Buchanan and Dr. Debbie Dailey share example scenarios to assist teachers in adapting their existing curriculum to include engineering practices. Drs. Joyce VanTassel-Baska and Bronwyn MacFarlane offer suggestions on how to create engineering problem-based learning lessons. To check for student progress and to guide instruction, Dr. Ann Robinson, Ms. Kristy Kidd, and Dr. Jill Adelson examine approaches and assessments for measuring student outcomes in engineering.

Section 4 addresses teacher professional development and student identification considerations when implementing engineering in K–8 classrooms. Dr. Alicia Cotabish, Dr. Umadevi Garimella, and Ms. Gina Howes Boshears examine the best methods for supporting teachers in implementing engineering in their classrooms through professional development. Additionally, Dr. Kinnaria Atit, Ms. Kay E. Ramey, Dr. David H. Uttal, and Dr. Paula Olszewski-Kubilius discuss the advantages of introducing engineering to students early on, examine specific skills that are necessary for engineering, provide avenues for integrating and differentiating engineering into K–8 curriculum, and consider the value of adding spatial skills to identify academically talented students.

The book concludes with Appendix A, which provides resources for educators, parents, and students, and Appendix B, which provides descriptions of engineering in both formal and informal environments.

We thank our esteemed colleagues, who diligently provided their expertise and knowledge in writing their chapters; the reviewers, who provided additional eyes to help ensure we produced a quality product; NAGC, which is always supportive in providing resources for educators; and the general education and gifted classroom teacher—for your time and effort in educating our children.

REFERENCES

Achieve, Inc. (2014). *Next Generation Science Standards*. Washington, DC: Author.

Change the Equation. (2012). STEM help wanted. *Vital signs: Reports on the condition of STEM learning in the U.S.* Retrieved from http://changetheequation. org/sites/default/files/ CTEq_VitalSigns_Supply (2).pdf

Dailey, D. (2015). Elementary science curriculum for gifted learners. In B. MacFarlane (Ed.), *STEM education for high-ability learners: Designing and implementing programing* (pp. 17–32). Waco, TX: Prufrock Press.

National Research Council. (2012). *A framework for K–12 science education: Practices, crosscutting concepts, and core ideas*. Washington, DC: The National Academies Press.

NGSS Lead States. (2013). *Next generation science standards: For states, by states*. Washington, DC: The National Academies Press.

Rothwell, J. (2013). The hidden STEM economy. *Brookings*. Retrieved from http:// www.brookings.edu/research/reports/2013/06/10-stem-economy-rothwell

VanTassel-Baska, J. (1992). *Planning effective curriculum for gifted learners*. Denver, CO: Love.

VanTassel-Baska, J. (1998). Planning science programs for high ability learners. *ERIC Clearinghouse on Disabilities and Gifted Education*. Retrieved from http://www.ericdigests.org/1999-3/science.htm

An Overview of Engineering/Technology Standards and Practices

Cheryll M. Adams

Currently, there are no standalone national engineering standards in K–12 education, although this issue has been studied and debated for quite some time as a way to demonstrate that engineering is an established component of STEM (National Academy of Engineering, The Committee on Standards for K–12 Engineering Education [NAE], 2010; National Research Council [NRC], 2009). The purpose of this chapter is to set the discussion of engineering standards first within the context of STEM standards, look at the pros and cons of having national engineering standards, examine what standards in engineering currently exist, identify by which criteria the current standards may be assessed, and explore current best practices using the available standards.

A (BRIEF) HISTORY OF STANDARDS IN STEM

Having educational standards is not a new idea, but most standards from the 1800s through the early 1900s were in the form of policies and committee reports that reflect what we currently call standards (NAE, 2010). The current focus on standards resulted from the publication of the National Commission on Excellence in Education's 1983 *A Nation at Risk* and from President George H.W. Bush and a group of governors, whose 1989 educational summit "laid the groundwork for the Goals 2000 Education Project" (NAE, 2010, p. 7). At that time, the National Association for Teachers of Mathematics and the American Association for the Advancement of Science published documents in support of national standards in math and science, respectively, followed in 1996 by the development of the National Science Education Standards by the National Research Council (NAE, 2010). From this beginning focus, we now have the Common Core State Standards in Mathematics (CCSSM; National Governors Association Center for Best Practices [NGA], & Council of Chief State School Officers [CCSSO], 2010) and the Next Generation Science Standards (NGSS; NGSS Lead States, 2013). (For a detailed chronology of STEM-related standards initiatives in the past 40 years, see NAE, 2010, pp. 12–14.)

Both the CCSSM and the NGSS have resulted in a restructuring of STEM education as schools address the new standards. Unfortunately, the "T" and particularly the "E" in STEM have not been in the spotlight as much as the "S" and "M." Technology education (the "T") has a long history (including industrial and manual arts): a set of standards specifying what students should know, understand, and be able to do to be considered technologically literate; and a core group of teachers who teach classes in technology (International Technology Education Association [ITEA], 2000). These standards include some engineering-related learning goals, and based on the shift in technology education toward engineering, in 2010, ITEA changed its name to the International Technology and Engineering Educators Association (ITEEA). According to the NRC (2009),

> In contrast to science, mathematics, and even technology education, all of which have established learning standards and a long history in the K–12 curriculum, the teaching of engineering in elementary and secondary schools is still very much a work in progress. Not only have no learning standards been developed, little is available in the way of guidance for teacher professional development, and no national or state-level assessments of student accomplishment have been developed. In addition, no single organization or central clearinghouse collects information on K–12 engineering education. (p. 2)

Furthermore, since the early 1990s, less than 10% of K–12 students have had experience with formal engineering curricula (NRC, 2009). Clearly engineering seems to have been the stepchild of STEM education. However, engineering is currently getting some well-deserved attention, as the debate over whether or not to have national engineering standards moves front and center.

ENGINEERING STANDARDS: TO BE OR NOT TO BE?

Although the number of K–12 students being exposed to engineering-related materials is small, a few studies indicate that engineering education can "stimulate interest and improve learning in mathematics and science as well as improve understanding of engineering and technology" (NAE, 2010, p. 1). Thus, the Committee on Standards for K–12 Engineering Education (NAE, 2010) sought to determine whether engineering standards might "improve the quality and increase the amount of teaching and learning of engineering in K–12 education" (p. 1). The committee agreed that it is possible through careful deliberation to agree on a set of K–12 standards that would reflect the core content of K–12 engineering and inform the development of curricula with the possibility of giving engineering its own identity as a separate discipline. After considerable research and study, the committee concluded that, while in theory it is possible to develop such standards, in practicality, it would be challenging to ensure the standards would be useful and implemented effectively. This conclusion was based on four issues:

> (1) there is relatively limited experience with K–12 engineering education in U.S. elementary and secondary schools, (2) there is not at present a critical mass of teachers qualified to deliver engineering instruction, (3) evidence regarding the impact of standards-based educational reforms on student learning in other subjects, such as mathematics and science, is inconclusive, and (4) there are significant barriers to introducing stand-alone standards for an entirely new content area in a curriculum already burdened with learning goals in more established domains of study. (NAE, 2010, pp. 1, 18, 19)

Two alternatives to developing standalone engineering standards—infusion and mapping—have been examined by researchers. Infusion is defined as "including the learning goals of one discipline—in this case engineering—in educational standards for another discipline" (NAE, 2010, p. 23). Mapping means "drawing attention explicitly to how and 'where' core ideas from one discipline relate to the content of existing standards in another discipline" (p. 28). Thus, infusion is proactive while mapping is retrospective.

Engineering standards are infused in two sets of standards in the STEM fields, the NGSS (NGSS Lead States, 2013) and the Standards for Technological Literacy (STL; ITEA, 2000). Three of the 20 STL are focused specifically on engineering: Standard 8: Students will develop an understanding of the attributes of design; Standard 9: Students will develop an understanding of engineering design; and Standard 11: Students will develop the abilities to apply the design process. Because the three engineering standards exist in the STL, I mention them here in passing, but because their potential to impact engineering education is small, there will be no further discussion of them.

In the NGSS, Engineering, Technology, and Applications of Science (ETS) is one of the four Disciplinary Core Ideas. Furthermore, Dimension 1 of the NGSS is "Science and Engineering Practices," which delineates the eight behaviors that scientists and engineers engage in as they participate in the processes of investigation or design, respectively. Interestingly, there is no mention of engineering in the CCSSM.

Although there are currently no instances of mapping engineering onto other national standards, some have suggested mathematics, history, civics, art, and college and career standards might be appropriate areas to consider (NAE, 2010). There are others who have offered actual guidelines for engineering standards (American Society for Engineering Education [ASEE], 2008a; Moore et al., 2014; Moore, Tank, Glancy, & Kersten, 2015). We will look at these guidelines later in this chapter. Thus, there is an agreement that standards in engineering education are important and needed (Brophy, Klein, Portsmore, & Rogers, 2008; Chandler, Fontenot, & Tate, 2011; Moore et al., 2014; Moore et al., 2015; NRC, 2009); the issue is how to manage adding more standards without "perpetuat[ing] the politics and territorial disputes among the science, technology, engineering, and mathematics disciplines" (Bybee, 2010. p. 11).

ENGINEERING STANDARDS WITHIN THE NGSS

Because engineering standards have been infused into the NGSS to a greater extent than they have been in any other set of standards, this section will examine some of the pros and cons of this approach. The NGSS represent a shift in science education from the typically "inch deep and mile wide" coverage of topics to an in-depth study of major concepts. As previously mentioned, another shift is placing engineering front and center on equal footing with life science, physical science, and Earth and space science, increasing the visibility and understanding of engineering by "raising engineering design to the same level as scientific inquiry in science classroom instruction at all levels, and by emphasizing the core ideas of engineering design and technology applications" (NGSS Lead States, 2013, Executive Summary, p. 1). In support of this shift, Moore et al. (2015) provide three main arguments for the teaching of engineering in K–12 learning:

> (1) engineering thinking helps with the development of 21st century skills in students, (2) engineering pedagogies have potential to increase student achievement in mathematics and science, and (3) engineering contexts have potential to increase student interest in STEM disciplines and careers. (p. 298)

However, the infusion approach in the NGSS includes only engineering design; as the developers state, "It is important to point out that the NGSS do not put forward a full set of standards for engineering education, but rather include only practices and ideas about engineering design that are considered necessary for literate citizens" (NGSS Lead States, 2013, Appendix I, p. 3). So how is engineering infused into the NGSS, and is there enough? A simple count of the Performance Expectations (PEs) finds 42 of 208 of those addressing engineering design. A review of the NGSS by Moore and colleagues (2015) found an additional seven PEs not identified explicitly as engineering by the developers. They noted that there were 76 Learning Goals (LGs) associated with the 49 PEs. They affirm that the NGSS is currently the best attempt to address engineering standards at the national level, but after careful study using the framework presented in the next section, they still find the NGSS lacking. Specifically the third- to fifth-grade band had no coverage of engineering thinking or habits of mind, limited discussion of engineering ethics, and nowhere was there coverage of teamwork; they see these components as central to any study of engineering. They stress that the most comprehensive coverage of engineering in the NGSS comes from using both the PEs and the associated LGs.

GUIDELINES FOR K-12 ENGINEERING STANDARDS

There have been two recent attempts at establishing guidelines for engineering standards: *A Framework for a Quality K–12 Engineering Education* (Moore et al., 2014; Moore et al., 2015), and the *K–12 STEM Guidelines for All Americans* (ASEE, 2008a).

Framework for a Quality K-12 Engineering Education

One promising attempt at delineating the essential elements of a K–12 engineering education is the *Framework for a Quality K–12 Engineering Education* (Moore et al., 2014; Moore et al., 2015). The reason for designing the framework, according to Moore and colleagues (2014) was

> to address the need for a clear definition of engineering at the elementary and secondary levels through the development of a framework for describing, creating, and evaluating engineering in K–12 settings. Such a definition could help to guide the development of robust engineering and STEM education initiatives and inquiries into their effectiveness. (p. 2)

Using a design-research paradigm, the researchers sought to answer the question, "What constitutes a quality and comprehensive engineering education at the K–12 level?" (p. 2). According to Moore et al. (2014),

> the indicators were determined based on an extensive literature review, established criteria for undergraduate engineering programs and professional organizations, document content analysis of state academic standards, evaluation of classroom practice and curriculum implementation, and in consultation with experts in the fields of engineering and engineering education. (p. 2)

The researchers indicate the framework may be used to evaluate the degree to which "academic standards, curricula, and teaching practices address the important components of a quality K–12 engineering education" and "to inform the development and structure of future K–12 engineering and STEM education standards and initiatives" (p. 2). Within the current framework are 12 key indicators that correspond to the essential elements based on this process. Moore et al. (2015) identified the key indicators as: Complete Processes of Design (POD); Problem and Background (POD–PB); Plan and Implement (POD–PI); Test and Evaluate (POD–TE); Apply Science, Engineering, Mathematics Knowledge (SEM); Engineering Thinking (EThink); Conceptions of Engineers and Engineering (CEE); Engineering Tools, Techniques, and Processes (ETool); Issues, Solutions, and Impacts (ISI); Ethics (Ethics); Teamwork (Team); and Communication Related to Engineering (Comm–Engr). The framework went through an extensive process of review, testing under controlled conditions, feedback, and revisions. The current framework is iteration five.

Based on the framework, Moore et al. (2015) delineated four criteria that must be met before a set of standards can be acknowledged as providing a comprehensive engineering education:

> (1) the standards must explicitly focus on engineering; (2) Processes of Design [POD], Applications of Science, Engineering, and Mathematics [SEM], and Engineering Thinking [EThink]) must be included multiple times in all four grade bands; (3) Conceptions of Engineers and Engineering (CEE), and Engineering Tools and Processes (ETool) should be present multiple times in at least two of the grade bands; and (4) Issues Solutions & Impacts [ISI], Ethics, and Engineering Communication [Comm-Engr]) must be addressed multiple times in at least one of the grade bands. (p. 305)

The framework has been used to assess the degree to which science standards in each of the 50 states include engineering (Moore, Tank, Glancy, Kersten, & Ntow, 2013; Moore et al., 2015), to examine the inclusion of engineering in the NGSS (Moore et al., 2015), and to assess the degree to which integrated science and engineering curriculum align with the framework (Glancy et al., 2014). (For a complete discussion of the framework and its development see Moore et al., 2014, 2015.)

K-12 STEM Guidelines for All Americans

The Corporate Member Council (CMC) of the ASEE represents the interests of corporations, government agencies, and nonprofit organizations that are interested in engineering education. Previously, the CMC focused on engineering at the collegiate level. A special interest group (SIG) of the CMC, the K–12 Science, Engineering, and Technology Workforce, met several times, as indicated by documents on its website, to discern a set of guidelines for engineering education. These guidelines are referred to as *K–12 STEM Guidelines for All Americans* or *Engineering Activities Are for All Americans*, depending on the document or section of the website that is accessed. For convenience, I will refer to them as the CMC SIG Guidelines.

According to the rationale for developing the CMC SIG Guidelines (ASEE, 2008b), engineering could be considered the umbrella under which all STEM courses could be taught:

INTRODUCTION

It is a vehicle to bring rigor, relevance and context to the teaching of the other three subjects in an integrated manner. Using engineering activities as a vehicle allows core subjects to be taught more efficiently, in a way that leads to increased retention, to the ability to apply diverse knowledge and concepts to different situations, to synthesis, creativity and problem solving . . . all vital to the necessary 21st century skills. (p. 1)

The CMC noted that many students dislike math and science because they do not see how it relates to their own lives and futures. The committee advocated using engineering in grades K–12 as

a stealth approach to reaching children that haven't and aren't being reached in the current methods of teaching of isolated subjects. Using engineering activities in the classroom can have the ultimate result, where more students learn more and understand the concepts better (ASEE, 2008b, p. 1).

The CMC stressed that the engineering approach aligns with Bloom's taxonomy, although the committee used Bloom's 1956 version, not the revised version (Anderson & Krathwohl, 2001), which seems to be a better fit.

The CMC proposed the CMC SIG Guidelines as a way to help the K–12 educational system improve student performance. These guidelines are for curriculum writers and educators to use to "emphasize the characteristics and qualities of a STEM event that should be fostered" (ASEE, 2008b, p. 7). The intent of the guidelines is to provide objectives but allow teachers leeway in how they interpret those when developing activities. The CMC stressed that "the Engineering Activity will be the meaningful event to support the math, science, or technical concept in the process" (ASEE, 2008b, p. 7). By using the CMC SIG Guidelines, teachers should be able to prepare students for a variety of positions in engineering.

The CMC SIG Guidelines consist of five dimensions, each of which includes what students should understand and be able to do. The dimensions are (1) Engineering Design; (2) Connecting Engineering to Science, Technology and Mathematics; (3) Nature of Engineering; (4) Communication and Teamwork; and (5) Engineering and Society. For example, Dimension 2 lists "technological literacy" and "how things work" as foci. Students should understand "the essential concepts of and how to apply science, technology, and mathematics as they pertain to engineering," and be able to "apply concepts of science, technology, and mathematics to engineering processes and problems" (ASEE, 2008a, p. 2). The broad understanding and procedural guidelines are further subdivided. (For the entire set of guidelines, see https://www.asee.org/member-resources/councils-

and-chapters/corporate-member-council/special-interest-group/cmc-k12-stem-guidelines-for-all-americans.pdf.)

The rationale of the CMC SIG mentioned that the committee would like to have the education community review the guidelines and offer feedback, so that the guidelines can be revised. Furthermore, it mentioned plans to assemble an expert committee to review the guidelines for improvement (ASEE, 2008b). According to the plan, this committee would meet each year for the next 4 years. This panel review would have begun in 2009 or 2010 and continued to 2012 or 2013, but currently there is no further mention of reviews or revised guidelines at the ASEE website or in the literature.

NEXT STEPS

In 2009, the Committee on Understanding and Improving K–12 Engineering Education in the United States set forth three general principles for K–12 engineering education (NRC, 2009):

- **Principle 1.** K–12 engineering education should emphasize engineering design.
- **Principle 2.** K–12 engineering education should incorporate important and developmentally appropriate mathematics, science, and technology knowledge and skills.
- **Principle 3.** K–12 engineering education should promote engineering habits of mind. (pp. 151–152)

Although the literature suggests there are benefits to teaching engineering to K–12 students (e.g., improved performance in science and mathematics, an increase in the number of students who pursue careers in engineering), the data are limited to support these claims (NRC, 2009). A larger body of high-quality outcomes data is needed before engineering education will be a focus of K–12 education (NRC, 2009). To that end, inroads toward providing guidelines for including engineering in K–12 education have been made through the development of the NGSS (NGSS Lead States, 2013), *A Framework for a Quality K–12 Engineering Education* (Moore et al., 2014), and the *K–12 STEM Guidelines for All Americans* (ASEE, 2008a). An increasing body of research surrounds *A Framework for a Quality K–12 Engineering Education*, focusing on evaluation of state standards and the NGSS and curriculum development. We have an ideal opportunity to work toward expanding the teaching of engineering in grades K–12 by using these evidence-based tools and guidelines to develop engineering standards. Like the NGSS and the CCSSM, the engineering guidelines are intended for use with all students. Whether engineering standards are subsequently developed as stand-alone or remain infused into other standards such as the NGSS, they will still need to be differentiated to meet the needs of gifted and advanced learners.

REFERENCES

American Society for Engineering Education, Corporate Member Council. (2008a). *K–12 STEM guidelines for all Americans.* Retrieved from https://www.asee.org/member-resources/councils-and-chapters/corporate-member-council/special-interest-group/cmc-k12-stem-guidelines-for-all-americans.pdf

American Society for Engineering Education, Corporate Member Council. (2008b). *K–12 STEM guidelines rationale for all Americans.* Retrieved from https://www.asee.org/member-resources/councils-and-chapters/corporate-member-council/special-interest-group/cmc-k12-stem-guidelines-for-all-americans-rationale.pdf

Anderson, L., & Krathwohl, D. (Eds.). (2001). *A taxonomy for learning, teaching, and assessing: A revision of Bloom's taxonomy of educational objectives* (Complete ed.). New York, NY: Longman.

Bloom, B. (Ed.). (1956). *Taxonomy of educational objectives: The classification of educational goals. Handbook I: Cognitive domain.* New York, NY: Longmans Green.

Brophy, S., Klein, S., Portsmore, M., & Rogers, C. (2008). Advancing engineering education in P–12 classrooms. *Journal of Engineering Education, 97,* 369–387.

Bybee, R. W. (2010). K–12 engineering education standards: Opportunities and barriers. In National Research Council (Ed.), *Standards for K–12 engineering education?* (pp. 55–66). Washington, DC: The National Academies Press.

Chandler, J., Fontenot, A. D., & Tate, D. (2011). Problems associated with a lack of cohesive policy in K–12 pre-college engineering. *Journal of Pre-College Engineering Education Research, 1*(1), 40–48.

Glancy, A. W., Moore, T. J., Guzey, S. S., Mathis, C. A., Tank, K. M., & Siverling, E. A. (2014, June). *Examination of integrated STEM curricula as a means toward quality K–12 engineering education (Research to practice).* Paper presented at the meeting of the American Society for Engineering Education Annual Conference and Exposition, Indianapolis, IN.

International Technology Education Association. (2000). *Standards for technological literacy: Content for the study of technology.* Reston, VA: Author.

Moore, T. J., Glancy, A. W., Tank, K. M., Kersten, J. A., Smith, K. A., & Stohlmann, M. S. (2014). A framework for quality K–12 engineering education: Research and development. *Journal of Pre-College Engineering Education Research, 4*(1), 1–13.

Moore, T. J., Tank, K. M., Glancy, A. W., & Kersten, J. A. (2015). NGSS and the landscape of engineering in K–12 state science standards. *Journal of Research in Science Teaching, 52,* 296–318.

Moore, T. J., Tank, K. M., Glancy, A.W., Kersten, J. A., & Ntow, F. D. (2013, June). *The status of engineering in the current K–12 state science standards (Research to practice).* Paper presented at the American Society for Engineering Education Annual Conference and Exposition, Atlanta, GA.

National Academy of Engineering, Committee on Standards for K–12 Engineering Education. (2010). *Standards for K–12 engineering education?* Washington, DC: The National Academies Press.

National Commission on Excellence in Education. (1983). *A nation at risk: The imperative for educational reform.* Washington, DC: U.S. Government Printing Office.

National Governors Association Center for Best Practices, & Council of Chief State School Officers. (2010). *Common Core State Standards for mathematics.* Washington, DC: Author.

National Research Council. (1996). *National science education standards.* Washington, DC: The National Academies Press.

National Research Council. (2009). *Engineering in K–12 education: Understanding the status and improving the prospects.* Washington, DC: The National Academies Press.

NGSS Lead States. (2013). *Next Generation Science Standards: For states, by states.* Washington, DC: The National Academies Press.

SECTION 1

Key Components of Engineering Instruction for K-8 High-Ability Learners

DOI: 10.4324/9781003234951-2

CHAPTER 1

Spatial Ability, Engineering, and Robotics for High-Ability Learners

Steve V. Coxon

Spatial ability is vitally important to engineers and can be taught to children through robotics. The importance of spatial ability to engineering and other STEM fields has been known since before the launch of Sputnik (Super & Bachrach, 1957), but from the sudden yet fleeting interest in STEM education that followed that launch to the present renewed interest, spatial ability has been almost entirely disregarded in K–12 schools (Webb, Lubinski, & Benbow, 2007). Coxon (2012) suggested that the end result of this lack of attention is a national shortage of college graduates with degrees in STEM fields, including engineering. Coxon (2013) offered many potential solutions, including early education in architecture, the arts, computer science, geographic information systems, geom-

 DOI: 10.4324/9781003234951-3

etry, physics, and, most notably, robotics engineering. Although the U.S. Bureau of Labor Statistics predicts STEM occupations will grow by 17% this decade, far too few high school graduates are pursuing STEM majors (Langdon, McKittrick, Beede, Khan, & Doms, 2011). Although 28% of college students begin as STEM majors, about half will either switch majors or drop out of school before graduating (Chen & Soldner, 2013). Given the demand for highly educated people in engineering fields such as robotics, coupled with the fact that engineers earn considerably more than the U.S. average, it may be surprising that few people choose to pursue engineering and other STEM fields.

Multiple national reports have called for increased attention on STEM education in grades K–12. These include recommendations for earlier interventions, to focus on gifted children, and to stress the importance of spatial ability (Coxon, 2012). Among these, the National Science Board's (NSB) 2010 report, *Preparing the Next Generation of STEM Innovators*, noted that few efforts have "focused on raising the ceiling of achievement for our Nation's most talented and motivated students" (p. 4). The NSB further outlined key issues, which include placing a priority on early interventions and measuring and developing spatial ability in children. These reports have been unheeded by schools, perhaps because school policymakers, administrators, and teachers have an overall lack of understanding of what spatial ability is, its importance to STEM fields (including engineering), and how to develop spatial ability among students. Our schools have always been predicated upon developing talents in the three Rs—Reading, wRiting, and aRithmetic—that is, those subjects that draw upon verbal and math abilities. Schools tend to be resistant to change, even when evidence is overwhelming. Spatial ability is the third missing element and its absence can have negative implications for students whose engineering potential is not developed prior to college.

SPATIAL ABILITY

Spatial ability is "the ability to generate, retain, retrieve, and transform well-structured visual images" (Lohman, 1993, p. 1). Spatial ability is complex and has multiple facets that can be further understood as subabilities, each focused on different aspects of image processing: generation, storage, retrieval, and transformation (Lohman, 1993). Coxon (2012) provided a significant summary of spatial ability, including that Galton (1880) became the first to suggest that understanding spatial ability as a human difference may be important to education when he put forth that learners who utilize visualization might benefit from different instructional strategies than more verbally centered learners. More than 100 years later, Gardner's (1983) *Frames of Mind* popularized the construct along with other domains of ability for educators. Contemporary research continues to support Galton's observation of learning differences for the spatially-able, including that the spatially-able are more often creative (Kim & Coxon, 2016; Liben, 2009), more likely to be introverted (Lohman, 1993), much more likely to have hobbies (Humphreys, Lubinski, & Yao, 1993), and possibly more likely to have reading disabilities (Lohman, 1994; Mann, 2006). They are also considerably more likely to become engineers (Wai, Lubinski, & Benbow, 2009).

Measuring Spatial Ability

Spatial ability can be measured in children in several different ways, as further described in Chapter 13. One other means to measure spatial ability in high-ability children is the use of off-grade-level tests, which has a long history in math and verbal domains in talent searches. For example, the Project Talent Spatial Battery, a series of four tests, including 2-D and 3-D rotations, abstract spatial reasoning, and mechanical reasoning, was designed for high school students; the Revised Purdue Spatial-Visual Test: Visualization of Rotations test was designed for college students. However, Coxon (2012) and Coxon and Dorhman (2015) found the tests suitable to differentiate spatial ability among 9–14 year olds. Unfortunately, national norms for presecondary age ranges are not available for either test. Still, schools may find it useful to use either instrument with a population of children and then select the top 5%–10% or more for programs that seek to further the talent development of spatially-able students, such as robotics programs.

Gender Differences

Researchers of spatial ability frequently find significant gender differences favoring males. Coxon (2012) summarized these differences. Differences are so acute that Lohman (2005) suggested using this knowledge to determine if a reasoning ability test is also measuring spatial ability. That is, if boys score significantly higher on an item overall, it is probably measuring spatial ability in addition to reasoning. Many studies identified gender differences, but researchers generally suggest that both sexes can improve on spatial ability with treatment. In a thorough literature review, Spence, Yu, Feng, and Marshman (2009) found that women generally do not perform as well as men in spatial activities, especially mental rotation. In another literature review, Liu, Uttal, Marulis, and Newcombe (2008) reported that improvement can be made on spatial tasks with training, with those performing lowest initially making the largest gains after intervention. However, the study also revealed that, while improvement with intervention was made by both sexes, women generally did not perform as well as men before

KEY COMPONENTS OF ENGINEERING INSTRUCTION

or after intervention. Of course, while this applies in general, there are spatial-ly-able females. Females in the top 20% of spatial ability measures tend to be three times more likely to major in physical sciences, mathematics, engineering, and computer science as their high-verbal same-sex peers, while males are only twice as likely to major in those subjects as their high verbal-ability same-sex peers (Humphreys et al., 1993). As spatially-able females are proportionally more likely to go into the above fields, improvement in spatial ability may have an even stronger impact on female outcomes than male outcomes. It remains unknown if these differences are due to genetics, environmental factors such as the toys children are provided (e.g., LEGOs and other building toys are disproportionately given to boys), messages they receive, or biased instruments, but it appears likely that gender differences are a result of complex interactions of genetic and environmental factors (Coxon, 2012).

Spatial Ability and General Ability (*g*)

Spatial ability has long been considered a facet of *g*. General intelligence is a natural, largely in-born set of mental traits (Carroll, 1994; Jensen 1984). Spatial ability has been included in measures of intelligence from the first Stanford-Binet test, which included spatial items (Terman, 1916), to Project Talent in the 1960s (Wai et al., 2009), to some more recent assessments. Jensen (1984) found that general intelligence (*g*) generally accounts for all of the significantly predicted variance in job performance. However, he also found that in some cases, where jobs such as engineering require spatial ability, the overall validity for predicting job performance is significantly improved by tests of spatial visualization in addition to measures of *g*. Humphreys and colleagues (1993) suggested that while the level of ability is related to *g*, the patterns of educational and occupational choices are related to group factors, such as spatial ability. Carroll (1994) theorized that while more than half of various tests can be explained by *g*, the rest of the variance is explained by a few lower order factors including spatial ability. Although genetics are a primary factor in determining intelligence (Herrnstein & Murray, 1994), the aforementioned research and more suggests that performance on measures of spatial ability is improvable through treatment (Coxon, 2012; Kim & Coxon, 2016; Liu et al., 2008; Lohman, 1993). Thus, it appears possible to serve the spatially-able in schools by developing their talents through challenging spatial experiences such as robotics programs.

SPATIAL ABILITY IN SCHOOLS

Despite the potential for student improvement, spatial talents are rarely developed in K–12 schools (Coxon, 2012). The NSB (2010) noted that "individuals with spatial abilities are routinely overlooked because spatial ability is rarely measured" (p. 13). The board suggested that this group is underserved in schools and "an untapped pool of talent critical for our highly technological society" (p. 13). Spatial ability in childhood is a reliable predictor of a STEM career as an adult, and students with high spatial ability are particularly well-suited for engineering careers (Wai et al., 2009; Webb et al., 2007).

Wai and colleagues (2009) summarized longitudinal research on spatial ability in Project Talent. Project Talent gave 40 tests, including four spatial, to 377,000 high school students in the early 1960s and followed them for 50 years, providing multiple outcome data. The study demonstrated the importance of spatial ability to success in engineering and other STEM fields. In particular, 90% of participants who had earned a Ph.D. in a STEM field by the end of the 11-year follow-up possessed spatial ability in the top quartile; 45% of STEM doctorates were awarded to participants in the top 4% of spatial ability. Despite these and other similar, strong findings, Wai and colleagues found that more than half of participants in the top 1% of spatial ability were not within the top 3% of math and verbal ability, leaving them much less likely to qualify for gifted programs where spatial assessments are rarely used for identification. These students have little chance of receiving an education appropriate to their abilities. According to Mann (2006), students with spatial gifts tend to be undereducated and underemployed as adults, compared to students with similar gifts in mathematical and verbal areas. To address these educational deficiencies, engaging students in robotics has the potential to facilitate and develop spatial talents for engineering in K–12 students (Coxon, 2012).

Spatial Ability, Engineering, and Robotics

The small amount of research on robotics education shows promise for improving measured spatial ability. Coxon (2012) investigated spatial ability among 75 students (ages 9–14) in one of the few available quantitative studies of spatial ability among K–12 students. In the controlled experimental study of a FIRST LEGO League (FLL) robotics competition simulation, Coxon found significant and meaningful gains for boys, but not for girls. This may be due to differences in prior experiences with spatial toys such as blocks and LEGOs, genetic differences, or problems with the test used. Other studies found similar results, including improvement in spatial ability and spatially related engineering concepts, among middle and high school students (Verner, 2004) and undergraduate sophomore and junior level students (Wang, LaCombe, & Rogers, 2004). Petre and Price (2004) interviewed participants and their coaches engaged in robotics competitions, including FLL competitions, and determined that robotics worked effectively as an engaging vehicle to guide children toward an understanding of programming and engineering principles, and that this learning was generalizable to other programming and engineering situations. One pathway for engagement in robotics engineering for children are the FIRST competitions.

FIRST Robotics Programs

FIRST is the Foundation for Inspiration and Recognition in Science and Technology. The nonprofit organization designs programs to encourage young people to pursue education in STEM fields. FIRST (2016) reported that 41% of high school participants go on to major in engineering, including 33% of female participants. The foundation sponsors the FIRST LEGO League for ages 9–14, as well as the Junior FLL for ages 6–9 years and two high school level robotics programs, FIRST Tech Challenge and FIRST Robotics Competition. Together, these programs served more than 400,000 children in more than 40 countries in 2015 (FIRST, 2016). In FLL, for example, a real-world STEM topic is chosen. Topics have included nanotechnology, medical innovation, energy production, and global climate change. In the FLL, students are asked to design, engineer, and program a LEGO robot to perform tasks on a table set up with LEGO creations related to the year's real-world science theme, as well as to complete a research project based on that theme involving field experiences. For example, when the theme was energy production, students could design, engineer, and program a LEGO robot to place a LEGO solar panel on a LEGO house among other challenges, and they could research energy-saving measures, complete an energy audit on a local government building, and present creative cost-saving recommendations to local government officials (Coxon, 2009). To effectively implement FIRST programs and other forms of robotics education, teachers and instructors need considerable professional development and ongoing support.

PROFESSIONAL DEVELOPMENT

Coxon (2013) offered several recommendations for professional development on developing spatial talents. It is important that teachers become familiar with the special characteristics of the spatially-able as well as the possibilities for challenging them, particularly in engineering. Professional development could be focused solely on the spatially able, or as a major facet of professional development on gifted learners more generally, differentiation for all learners, or engineering education. Although a workshop is a good beginning, professional development must be ongoing to be successful in the long term. With effective professional development and support, teachers can successfully implement robotics into their classrooms. For example, Coxon and Dohrman (2015) reported teachers were able to successfully implement robotics curriculum in their home schools after undergoing 25 hours of professional development on robotics working directly with small groups of students during a summer program under guidance from robotics education experts. Teachers must also be provided time to prepare lessons and provided materials for classroom use—coupled with funding to support the acquisition of necessary instructional materials, such as robotics sets.

CURRICULAR POSSIBILITIES

Robotics engineering can be incorporated into math and science curriculum. For example, Coxon and Dohrman (2015) reported on the Children using Robotics for Engineering, Science, Technology, and Math (CREST-M) project. CREST-M involved the creation of math curriculum involving LEGO WeDo robotics. In a controlled, experimental evaluation with 75 rising third- and fourth-grade students, researchers found significant and meaningful gains in math measurement skills. Students can utilize an engineering design loop in the creation of working robots tied to math standards. A design loop guides students through problem identification, research and ideation, convergence on a likely solution, a plan, a prototype, and testing. As a loop, students continue cycling through the process to make their product increasingly refined. For example, students might be asked to build a robot that can travel a specified distance or launch a ball to a specified height. Such work may simultaneously improve spatial ability if enough time and opportunities are offered to students.

CONCLUSION

Well-developed spatial ability improves STEM outcomes, including success in college engineering programs. To increase engineering potential, K–12 schools should focus on the development of spatial ability just as seriously as math and verbal domains, including finding spatially-gifted students and providing spatial curriculum appropriate to their abilities. Coxon (2013) offered many potential solutions, including early education in architecture, the arts, computer science, geographic information systems, geometry, physics, and, most notably, robotics engineering. Robotics curriculum and programs are one notable way in which spatial talent development can be accomplished, including for gifted students, as there is no ceiling for excellence. Teachers require significant professional development in order to be prepared to use robotics materials and implement robotics curriculum. Robotics can be integrated into core subject areas, such as math and science, to simultaneously encourage spatial growth while meeting standards. Preparing gifted students spatially will likely increase those ready for and interested in STEM fields, especially engineering, in their postsecondary education and beyond.

RESOURCES

Coxon, S. V. (2013). *Serving visual-spatial learners.* Waco, TX: Prufrock Press.

FIRST. (n.d.). *K–12 programs for kids.* Retrieved from http://www.firstinspires. org

Golon, A. (2008). *Visual-spatial learners: Differentiation strategies for creating a successful classroom.* Waco, TX: Prufrock Press.

Johnson, D. (2008). *Spatial reasoning: A mathematics unit for high-ability learners* (grades 2–4). Waco, TX: Prufrock Press.

Maker Shed. (n.d.). *Robot kits.* Retrieved from http://www.makershed.com/ collections/robot-kits

National Aeronautics and Space Administration. (n.d.). *Robotics for educators.* Retrieved from http://www.nasa.gov/audience/foreducators/robotics/home

REFERENCES

Carroll, J. B. (1994). An alternative, Thurstonian view of intelligence. *Psychological Inquiry, 5,* 195–197.

Chen, X., & Soldner, M. (2013). *STEM attrition: College students' paths into and out of STEM fields: Statistical analysis report.* Washington, DC: National

Center for Educational Statistics. Retrieved from http://nces.ed.gov/pubs 2014/2014001rev.pdf

Coxon, S. V. (2009). Challenging the neglected spatially gifted students with FIRST LEGO League. In B. MacFarlane & T. Stambaugh (Eds.), *An addendum to leading change in gifted education* (pp. 25–29). Williamsburg, VA: Center for Gifted Education.

Coxon, S. V. (2012). The malleability of spatial ability under treatment of a FIRST LEGO League simulation. *Journal for the Education of the Gifted, 35,* 291–316.

Coxon, S. V. (2013). *Serving visual-spatial learners.* Waco, TX: Prufrock Press.

Coxon, S. V., & Dohrman, R. L. (2015, October). *CREST-M curriculum: Children using robotics for science, technology, engineering, and math.* Paper presented at the Gifted Association of Missouri Conference, Columbia, MO.

FIRST. (2016). *At a glance.* Retrieved from http://www.firstinspires.org/about/at-a-glance

Galton, F. (1880). Statistics of mental imagery. *Mind, 5,* 300–318.

Gardner, H. (1983). *Frames of mind: The theory of multiple intelligences.* New York, NY: Basic Books.

Herrnstein, R. J., & Murray, C. (1994). *The bell curve: Intelligence and class structure in American life.* New York, NY: Free Press.

Humphreys, L. G., Lubinski, D., & Yao, G. (1993). Utility of predicting group membership and the role of spatial visualization in becoming an engineer, physical scientist, or artist. *Journal of Applied Psychology, 78,* 250–261.

Jensen, A. R. (1984). Test validity: g versus the specificity doctrine. *Journal of Social and Biological Structures, 7,* 93–18.

Kim, K. H., & Coxon, S. V. (2016). Fostering creativity using robotics among students in STEM fields to reverse the creativity crisis. In M. K. Demetrikopoulos & J. L. Pecore (Eds.), *Interplay of creativity and giftedness in science* (pp. 351–356). Rotterdam, The Netherlands: Sense.

Langdon, D., McKittrick, G., Beede, D., Khan, B., & Doms, M. (2011, July). STEM: Good jobs now and for the future. *U.S. Department of Commerce Economics and Statistics Administration, Office of the Chief Economist, Issue Brief #03-11.* Retrieved from http://www.esa.doc.gov/sites/default/files/reports/documents/stemfinaljuly14_1.pdf

Liben, L. S. (2009). Giftedness during childhood: The spatial-graphic domain. In F. D. Horowitz, R. F. Subotnik, & D. J. Matthews (Eds.), *The development of giftedness and talent across the life span* (pp. 59–74). Washington, DC: American Psychological Association.

Liu, L. L., Uttal, D. H., Marulis, L. M., & Newcombe, N. S. (2008). *Training spatial skills: What works, for whom, why and for how long?* Paper presented at the 20th annual meeting of the Association for Psychological Science, Chicago, IL.

Lohman, D. F. (1993). *Spatial ability and g*. Paper presented at the Spearman Seminar, University of Plymouth, England.

Lohman, D. F. (1994). Spatially gifted, verbally inconvenienced. In N. Colangelo, S. G. Assouline, & D. L Ambroson (Eds.), *Talent development: Proceedings from the 1993 Henry B. and Jocelyn Wallace National Research Symposium on Talent Development* (Vol. 2, pp. 251–264). Dayton, OH: Psychology Press.

Lohman, D. F. (2005). The role of nonverbal ability tests in identifying academically gifted students: An aptitude perspective. *Gifted Child Quarterly, 49,* 111–138.

Mann, R. L. (2006). Effective teaching strategies for gifted/learning disabled students with spatial strengths. *Journal of Secondary Gifted Education, 17,* 112–121.

National Science Board. (2010). *Preparing the next generation of STEM innovators: Identifying and developing our nation's human capital.* Arlington, VA: National Science Foundation.

Petre, M., & Price, B. (2004). Using robotics to motivate 'back door' learning. *Education and Information Technologies, 9,* 147–158.

Spence, I., Yu, J. J., Feng, J., & Marshman, J. (2009). Women match men when learning a spatial skill. *Journal of Experimental Psychology: Learning, Memory, and Cognition, 35,* 1097–1103.

Super, D. E., & Bachrach, P. B. (1957). *Scientific careers and vocational development theory.* New York, NY: Teachers College Press.

Terman, L. M. (1916). *The measurement of intelligence.* Boston, MA: Houghton Mifflin.

Verner, I. M. (2004). Robot manipulations: A synergy of visualization, computation and action for spatial instruction. *International Journal of Computers for Mathematical Learning, 9,* 213–234.

Wai, J., Lubinski, D., & Benbow, C. P. (2009). Spatial ability for STEM domains: Aligning over 50 years of cumulative psychological knowledge solidifies its importance. *Journal of Educational Psychology, 101,* 817–835.

Wang, E., LaCombe, J., & Rogers, C. (2004). *Using LEGO bricks to conduct engineering experiments.* Proceedings of the 2004 American Society for Engineering Education Annual Conference and Exposition, Reno, NV.

Webb, R. M., Lubinski, D., & Benbow, C. P. (2007). Spatial ability: A neglected dimension in talent searches for intellectually precocious youth. *Journal of Educational Psychology, 99,* 397–420.

KEY COMPONENTS OF ENGINEERING INSTRUCTION

CHAPTER 2

Engineering Design and Gifted Pedagogy

Eric L. Mann and Rebecca L. Mann

One does not have to look far in educational literature for articles stressing the need for improvement in how science, technology, engineering, and mathematics (STEM disciplines) are taught. The White House Office of Science and Technology Policy's February 2016 white paper, *STEM for All: Ensuring High-Quality STEM Education Opportunities for All Students*, outlines policies and priorities in an effort to provide every student opportunities "to join the innovation economy, have the tools to solve our toughest challenges, and be active citizens in our increasingly technological world" (p. 1). One of the priorities is to improve STEM teaching and support active learning by engaging students in problem-solving activities, ranging from presenting problems for students to solve before receiving instruction to engaging students in original research. For those who have spent some time in gifted education, the recommendations should invoke déjà vu.

 DOI: 10.4324/9781003234951-4

At the 18th World Conference on Gifted and Talented Children, we hosted a session entitled *Engineering Teachable Moments: Employing Engineering Design Activities*, where we shared five papers that linked engineering design activities with many of the pedagogical practices found in gifted education programs.[1] Reactions to our session were positive, but concerns were voiced about how engineering topics might align with other content and skills that need to be covered and the ability of teachers and students to undertake engineering tasks in K–8 classrooms. Since then, curriculum programs have been developed, most notably the one at the Museum of Science, Boston, that have successfully introduced accessible engineering tasks as early as kindergarten. The Next Generation Science Standards (NGSS) include engineering design as a content area throughout grades K–12 curricula. In discussing the addition of engineering design in the NGSS, the authors (NGSS Lead States, 2013) wrote, "providing students a foundation in engineering design allows them to better engage in and aspire to solve the major societal and environmental challenges they will face in the decades ahead" (p. 103). Although connections to gifted education may not be explicitly identified here, we see links to both the types of learning experiences gifted children thrive on and the development of co-cognitive factors that influence talent development (Renzulli, Kehler, & Fogarty, 2006).

PRINCIPLES OF LEARNING—DESIGNING MEANINGFUL LEARNING EXPERIENCES

In his book, *Creating Innovators*, Tony Wagner (2012) lists many of the acknowledged innovators of the era who dropped out of college to pursue their ideas. He describes them with a phrase borrowed from Henry Rutgers, "schooling was interfering with their education" (p. 56). One measure of this interference may be students' self-reported interest in mathematics and science, which drops from more than 80% in fourth grade to less than 40% just 4 years later (National Center for Educational Statistics [NCES], 2003). Although the potential reasons for this are many, the drop in interest suggests a significant loss of students with gifts and talents in the STEM disciplines as they disengage from school-based STEM tasks. In an effort to interest more individuals in engineering, the National Academy of Engineering (NAE; 2002) looked at the public's understanding of engineering. As part of that study, two prevailing messages of K–12 engineering outreach programs were identified: "Math and science are fun," and "engineers are important and contribute to the quality of life, economy, environment" (p.

---

1   These papers were published in Mann, Mann, Strutz, Duncan, and Yoon (2011).

2). At a time when children are beginning to consider career options, trying to interest them in exploring engineering opportunities with these messages, while their school experiences suggest otherwise, is counterintuitive.

Just as we strive to offer our students differentiated learning experiences tailored to their interests and readiness, there are several different gifted education curriculum models that provide the framework for creating challenging, meaningful learning experiences. One that incorporates the best from several models is the Parallel Curriculum Model (PCM; Tomlinson et al., 2009). Delving into the model is beyond the scope of this chapter; however, there are several points in the theory and beliefs that underpin the PCM that are particularly relevant to preparing our children for the complex, technological world in which we live. In the opening chapter, the authors (Tomlinson et al., 2009) discussed the rational and guiding principles for the model. Curriculum should . . .

- help students grapple with complex and ambiguous issues and problems;
- provide students opportunities for original, creative, and practical work in the disciplines;
- help students uncover, recognize, and apply the significant and essential concepts and principles that explain the structure and workings of the discipline, human behavior, and our physical world (added: and our engineered/technological world);
- help students develop a sense of themselves as well as of their possibilities in the world in which they live; and
- be compelling and satisfying enough to encourage student to persist despite frustration and understand the importance of effort and collaboration. (p. 3)

Within these few lines, the authors captured many of the outcome objectives we now find in the Standards for Mathematical Practice of the Common Core State Standards for Mathematics (CCSSM; National Governors Association Center for Best Practices [NGA], & Council of Chief State School Officers [CCSSO], 2010), the Science and Engineering practices in the NGSS (NGSS Lead States, 2013), and the Learning and Innovation Skills advocated by the Partnership for 21st Century Learning (2015). The four parallels in the PCM—the core curriculum, the curriculum of connections, the curriculum of identity, and the curriculum of practice—offer opportunities for students to engage in learning that is personally meaningful as they progress from novices to practicing professionals.

The Enrichment Triad Model (Renzulli, 1977) structures curriculum activities into three different but interconnected types. Type I, General Exploratory Activities, introduces children to a wide variety of topics in various disciplines designed to pique interest and generate questions from the child that lead to further, deeper investigations. Type II activities consist of both the development of creative/critical thinking and problem-solving skills as well as how-to knowledge that will provide the child with the tools needed to conduct the investigations. Type III activities engage students in seeking solutions to real-world problems they have identified.

In their September 2015 article in *Science and Children,* Tejaswini and Wendell described a process that closely follows the Enrichment Triad Model. Working with an elementary school science specialist, they identified a problem area within the school community—the care of the school's garden. The concerns were shared with the students, and the students then worked to define a specific problem they could solve—a Type I activity. Type II activities followed as students collected data; explored materials, tools, and simple machines; and learned about the needs of the plants in the garden, all leading to designing and testing of prototypes of different solutions. Selecting and implementing the best approach resulted in a solution to a meaningful, real-world problem—the culminating Type III activity.

A similar approach is employed in the Engineering is Elementary (EiE) curriculum (Museum of Science, Boston, 2016), which is based in part on the beliefs that:

- Engaging students in hands-on, real-world engineering experiences can enliven math and science and other content areas and motivate students to learn math and science concepts by illustrating relevant applications.
- Engineering fosters problem-solving skills, including problem formulation, iteration, and testing of alternative solutions.
- Engineering embraces project-based learning, encompasses hands-on construction, and sharpens children's abilities to function in three dimensions—all skills that are important for prospering in the modern world. (Hester & Cunningham, 2007, p. 3)

In EiE units, students are introduced to a problem through the eyes of a child who uses engineering to solve a real-world problem. Students are then asked to solve a similar problem. The challenge creates a "just-in-time" learning environment where students can ask questions about the knowledge and skills they need to solve the problem—an active, inquiry-driven, rather than passive, receptive learning environment.

Although middle and high school technology teachers have incorporated engineering design concepts in the classrooms for decades, conversations about pre-K–12 engineering education are relatively new. For example, the American Society for Engineering Education's Pre-College Education division was formed in 2003 with a meager budget and several dedicated individuals. Online resources such as TeachEngineering (https://www.teachengineering.org) and eGFI: Dream Up the Future (http://teachers.egfi-k12.org) provide access to growing sources of standards-based engineering lessons plans and activities to help introduce engineering and engineering design to students.

MAKING A DIFFERENCE—CO-COGNITIVE FACTORS AND TALENT DEVELOPMENT

The development of creativity, pursuit of solutions to real-world problems, generation of innovative ideas, and ability to make connections between previously unconnected ideas are necessary for the development of engineering thinking. They are also cornerstones of gifted education pedagogy. The final products within the EiE curriculum are not full Type III activities (those familiar with the Enrichment Triad Model would call them Type II 1/2) but do give students the necessary experience with the engineering design process needed to move to a Type III product. Engineering Products in Community Service (EPICS) are found in many undergraduate programs designed to give their students practice in solving ill-defined, open-ended problems, and developing real-world experiences working with clients to define, communicate, and evaluate potential solutions. Likewise, EPICS High (https://engineering.purdue.edu/EPICSHS) offers similar opportunities to high school students, and the concepts are working their way down into the elementary grades with community-based engineering projects (Edutopia, 2013; Swenson & Portsmore, 2013; Tejaswin & Wendell, 2015).

In *Light Up Your Child's Mind*, Renzulli and Reis (2009) write about creative-productive learning. This kind of learning experience changes going to school from time spent (wasted?) to time engaged in passionate, active learning. "Creative-productive learning takes place when a youngster is intent on developing an original something, a product that he hopes will have a positive impact on an audience of some kind" (p. 12). Creating something that effects positive change may be especially appealing to a gifted child who is more sensitive to values and moral issues (Davis, Rimm, & Siegle, 2011), and is often deeply committed to righting wrongs in the world. This sensitivity to human concerns is one of the six co-cognitive factors in Renzulli's Operation Houndstooth theory that forms the background for his Three-Ring Conception of Giftedness (Renzulli et al., 2006). This characteristic of giftedness involves altruism and empathy and leads to a desire to take action to help others. The misperceptions about engineers—that they do math and science and do not engage with societal or community concerns—may be a factor in discouraging gifted children from pursuing engineering as a career.

Spatial ability is a predictor of success in engineering (Humphreys, Lubinski, & Yao, 1993; Kell & Lubinski, 2013; Mann, 2014; Wai, Lubinski, & Benbow, 2009). The value schools of engineering place on spatial reasoning is highlighted by the increasing number of universities and colleges that are implementing courses for first-year students targeted at improving spatial skills (e.g., Martn-Dorta, Saorn, & Contero, 2008; Onyancha, Derov, & Kinsey, 2009; Sorby & Baartmans, 2000). Gifted students with spatial strengths excel in this area without a need for a remedial spatial reasoning course as they have a natural affinity for engineering thinking. Mann (2014) identified learning preferences of students with spatial strengths. These preferences for holistic instruction, visual ideation, and innovation align with the engineering design process. Students with spatial strengths use a holistic approach for solving problems and thrive when engaged in hands-on, real-world experiences, such as those employed in the EiE curriculum. Visual ideation is a critical skill for professionals in the engineering field as they construct mental images to determine design and functionality of products and services. Students with spatial strengths use visualization strategies as they worked to master course content. Engineering would be a stagnant field without innovation, an area in which gifted students with spatial strengths are very comfortable. They would rather generate their own problem-solving procedure than use a scripted approach. They enjoy tackling the complex and ambiguous issues and problems described in the PCM (Tomlinson et al., 2009). The inquiry-based learning situations utilized in the engineering design process closely align with the learning preferences of highly spatial students.

Unfortunately, statistics indicate that very few of our gifted spatial learners are following their areas of strengths and pursuing careers in the STEM areas such as engineering (Young & Bae, 1997). Participation in engineering design activities at the K–12 level would provide gifted students with high spatial ability the opportunity to work in their area of strength and give them experiences that may motivate them to pursue careers in the engineering field in the future. The benefit of participation in engineering design activities would not be seen solely by students who have innate ability in the area of spatial reasoning. All students would benefit, as it is possible to increase spatial problem solving performance through instruction. Sorby and Baartsman (2000) conducted a 6-year longitudinal study that compared engineering students who were randomly placed in a spatial skills course to those who did not enroll in the course. Retention, grade point average, and successful completion of a graphics course were all higher for students who were instructed in spatial reasoning strategies.

PREPARING FUTURE PROBLEM SOLVERS

The 2012 Programme for International Student Assessment (PISA) included an assessment of students' skills in tackling real-life problems. In this assessment (OECD, 2014), creative problem solving competency was defined as

> an individual's ability to engage in cognitive processing to understand and resolve problem situations where a method of solution is not immediately available. It includes the willingness to engage with such situations in order to achieve one's potential as a constructive and reflective citizen. (p. 30)

An outcome of this assessment was that, on average, only 20% of students in participating countries could solve very straightforward problems if they had a familiar situation as a reference. A recommendation from the study was to empower students to solve problems within meaningful contexts. Often such contexts are messy, data is missing, assumptions need to be made, constraints exist, and promising solution paths turn out to be dead ends. Parents and teachers need the courage to support all students and especially gifted and talented learners in the pursuit of becoming practicing professionals. Encouragement to take intellectual risks, learn from mistakes, and communicate thoughts effectively is especially important when levels of frustration are high. These same skills are vital aspects of engineering design and only can be developed when we challenge beyond "textbooks [that] deliver example problems in step-by-step format—and teach students to look for the steps as opposed to thinking for themselves" (Hines, 2012, p. 41). Step-by-step instructions deny students the opportunity to explore, to make decisions, to be creative, and to engage in constructive dialogue as they work as practicing professionals to solve problems. Although many of our gifted students have perfectionist tendencies, Hines (2012), a practicing engineer and a professor at Tufts University, wrote,

> I know that a wrong answer is an inevitable result of my humanity, so I have to work according to a discipline that will allow me and my colleagues to catch my mistakes. Clear communication of my thought process is fundamental to this discipline. (p. 41)

Engineering design challenges offer the opportunity to "motivate and challenge students' fundamental understanding in context of a creative process" (Hines, 2012, p. 41), attributes that are lacking when the solution method is laid out as a sequence of steps to follow.

AN EXAMPLE ENGINEERING PROBLEM-SOLVING CHALLENGE

Access to clean, safe drinking water is taken for granted in most communities in the United States. At least that was true until recently. Lead contaminations in the water supplies for Flint, MI, and Jackson, MS, are current news stories. Health concerns for children living in those areas are discussed in multiple venues and schools, churches, and other organizations across the country are sending bottled water to help.

Although the concerns and challenges in Flint and Jackson are real and immediate, obtaining clean, safe drinking water is a global issue. The EiE curriculum has two units that have potential connections. In *Water, Water Everywhere: Designing Water Filters*, students test a variety of materials to see how well they remove contaminates from river water to provide a healthier environment for a turtle living there. In the process of designing a temporary shelter for a pet frog, students explore the properties of membranes that allow water to pass through at a controlled rate in *Just Passing Through: Designing Model Membranes*. These same kinds of systems, water filters, and use of membranes in reverse osmosis systems are also ways to reduce contaminates in drinking water. As exploratory, Type I activities these units could suggest a variety of community-based engineering projects (curriculum of practice/Type III products). Along the way, Type II activities that provide students information about the source of containments; processes, materials, risks, and associated costs involved in remediation; health risks; and a variety of other topics will be needed to provide multiple opportunities to meet the needs of the core curriculum. Students will explore connections between the various disciplines as they use their math and science knowledge and process skills to test various solutions and their language and communications skills to share the results. Connections between the communities they live in and other communities in other parts of the world are possible as they explore the differences in water sources and environmental conditions. For some, the questions they asked and the problems they find may resonate with their strengths and interests leading to future studies and potential vocations (curriculum of identity).

FINAL THOUGHTS

Gifted children are passionate about learning in their area of interest(s). For those whose strengths lie in one or more of the STEM disciplines, the opportunity to engage in engineering design activities offers a welcome change from step-by-step, problem-solving exercises that converge on one expected solution found in so many textbook- or worksheet-based curricula. Although all students should have the opportunity for inquiry-based engineering design activities, programs for gifted and talented students are especially well-suited for these types of activities, as the grade-level curriculum can be compacted by buying additional time for more in-depth explorations and iterations to improve the final products.

Seeking ways to interest more individuals in engineering, NAE (2008) commissioned a study to look at ways to "re-brand" the profession. Messages such as "must be good at math and science" and "connecting science to the real world" were the most often reported and also most often viewed as a barrier to studying engineering. Interestingly, while the same is true for medicine, there is no shortage of medical school applicants. Engineering is an optimistic and innovative profession that has a direct impact on the lives of people. Although math and science skills are needed tools, other characteristics of engineering such as creativity, collaboration, and communication are equally vital. For the child who is seeking the means to change the world, engineering offers a way.

RESOURCES

This is a brief list of curriculum programs and online resources. It is not a complete list but a starting point to explore ways to introduce students to engineering design activities.

- *Building Math for Common Core State Standards Series (Grades 6–8):* http://walch.com/Building-Math-for-Common-Core-State-Standards-3-Book-Series.html
- *Edutopia's Education Video Library:* http://www.edutopia.org/videos. A couple of our favorites are:
  - » *"Wetlands Watchers: Kids Care for Their Environment":* http://www.edutopia.org/wetland-watchers-service-learning-video
  - » *"How Design Thinking Can Empower Young People":* http://www.edutopia.org/is-school-enough-design-thinking-video

- *eGFI: Dream Up the Future:* http://teachers.egfi-k12.org
- *Engineering by Design, International Technology and Engineering Educators Association:* http://www.iteea.org/STEMCenter/EbD.aspx
- *Engineer Girl:* http://www.engineergirl.org
- *Engineering Education: Museum of Science Boston:* http://www.mos.org/eie

> » *Engineering is Elementary (Grades 1–5):* http://www.eie.org/eie-curri culum
> » *Engineering Adventures (Grades 3–5):* http://www.eie.org/engineer ing-adventures
> » *Engineering Everywhere (Grades 6–8):* http://www.eie.org/engineer ing-everywhere

- *National Science Digital Library:* https://nsdl.oercommons.org
- *PBS Learning Media* (search for "engineering design" and choose a grade level): http://www.pbslearningmedia.org
- *Project Lead The Way:* https://www.pltw.org
  > » *Launch (Grades K–5):* https://www.pltw.org/our-programs/pltw-launch
  > » *Gateway (Grades 6–8):* https://www.pltw.org/our-programs/pltw- gateway

- *TeachEngineering: Curriculum for K–12 Teachers:* https://www.teachengi neering.org

REFERENCES

Davis, G. A., Rimm, S. B., & Siegle, D. (2011). *Education of the gifted and tal- ented* (6th ed.). Essex, England: Pearson.

Edutopia. (2013, December). *How design thinking can empower young people* [Video file]. Retrieved from http://www.edutopia.org/is-school-enough- design-thinking-video

Hester, K., & Cunningham, C. M. (2007). *Engineering is elementary: An engineer- ing and technology curriculum for children.* Paper presented at the American Society for Engineering Education Annual Conference and Exposition, Honolulu, HI. Retrieved from http://eie.org/sites/default/files/research_ article/research_file/ac2007full8.pdf

Hines, E. M. (2012, August). Principles for engineering education: Part 3. *Structure, 40–41.* Retrieved from http://www.structuremag.org/wp-content/ uploads/D-EdIssues-Hines-Aug121.pdf

Humphreys, L. G., Lubinski, D., & Yao, G. (1993). Utility of predicting group membership and the role of spatial visualization in becoming an engineer, physical scientist, or artist. *Journal of Applied Psychology, 78,* 250–261.

Kell, H. J., & Lubinski, D. (2013). Spatial ability: A neglected talent in educa- tional and occupational settings. *Roeper Review, 35,* 219–230.

Mann, R. L. (2014). Patterns of response: A case study of elementary students with spatial strengths. *Roeper Review, 36,* 60–69.

Mann, E. L., Mann, R. L., Strutz, M., Duncan, D., & Yoon, S. Y. (2011). Integrating engineering into K–6 curriculum: Developing talent in the STEM disciplines. *Journal of Advanced Academics, 22,* 659–658. doi:10.1177/1932202X11415007

Martín-Dorta, N., Saorín, J. L., & Contero, M. (2008). Development of a fast remedial course to improve the spatial abilities of engineering students. *Journal of Engineering Education, 97,* 505–513. doi:10.1002/j.2168-9830.2008. tb00996.x

Museum of Science, Boston. (2016). The EiE curriculum. *Engineering is elementary.* Retrieved from http://www.eie.org/eie-curriculum

National Academy of Engineering. (2002). *Raising public awareness of engineering.* Washington, DC: The National Academies Press.

National Academy of Engineering. (2008). *Changing the conversation: Messages for improving the public understand of engineering.* Washington, DC: The National Academies Press.

National Center for Educational Statistics. (2003). *Comparative indicators of education in the United States and other G-8 countries: 2002* (NCES Pub No. 2003-26). Retrieved from http://nces.ed.gov/pubs2003/2003026.pdf

National Governors Association Center for Best Practices, & Council of Chief State School Officers. (2010). *Common Core State Standards for mathematics.* Washington, DC: Author.

NGSS Lead States. (2013). *Next generation science standards: For states, by states.* Washington, DC: The National Academies Press.

OECD. (2014). *PISA 2012 results: Creative problem solving: Students' skills in tackling real-life problems* (Vol. 5). Retrieved from http://www.oecd.org/pisa/keyfindings/pisa-2012-results-volume-v.htm

Onyancha, R. M., Derov, M., & Kinsey, B. L. (2009). Improvements in spatial ability as a result of targeted training and computer-aided design software use: Analyses of object geometries and rotation types. *Journal of Engineering Education, 98,* 157–167. doi:10.1002/j.2168-9830.2009.tb01014.x

Partnership for 21st Century Learning. (2015, May). *P21 framework definitions.* Washington, DC: Author. Retrieved from http://www.p21.org/storage/documents/docs/P21_Framework_Definitions_New_Logo_2015.pdf

Renzulli, J. S. (1977). *The Enrichment Triad Model: A guide for developing defensible programs for the gifted and talented.* Mansfield Center, CT: Creative Learning Press.

Renzulli, J. S., & Reis, S. M. (2009). *Light up your child's mind: Finding a unique pathway to happiness and success.* New York, NY: Little, Brown and Company.

Renzulli, J. S., Koehler, J. L, & Fogarty, E. A. (2006). Operation Houndstooth intervention: Social capital in today's schools. *Gifted Child Today, 29*(1), 14–24.

Sorby, S. A., & Baartmans, B. J. (2000). The development and assessment of a course for enhancing the 3-D spatial visualization skills of first year engineering students. *Journal of Engineering Education, 89,* 301–307. doi:10.1002/j.2168-9830.2000.tb00529.x

Swenson, J. E. S., & Portsmore, M. D. (2013, June). *Dynamics of 5th grade students engineering service learning projects.* Paper presented at the American Society for Engineering Education Annual Conference and Exposition, Atlanta, GA. Retrieved from https://peer.asee.org/dynamics-of-5th-grade-students-engineering-service-learning-projects

Tejaswini, D., & Wendell, K. (2015). Community-based engineering. *Science and Children, 53*(1), 67–73.

Tomlinson, C. A., Kaplan, S. N., Renzulli, J. S. Purcell, J. H., Leppien, J. H., Burns, D. E., . . . Imbeau, M. B. (2009). *The parallel curriculum: A design to develop learner potential and challenge advanced learners* (2nd ed.). Thousand Oaks, CA: Corwin Press.

Wagner, T. (2012). *Creating innovators: The making of young people who will change the world.* New York, NY: Scribner.

Wai, J., Lubinski, D., & Benbow, C. P. (2009). Spatial ability for STEM domains: Aligning over fifty years of cumulative psychological knowledge solidifies its importance. *Journal of Educational Psychology, 101,* 817–835.

White House Office of Science and Technology Policy. (2016, February). *STEM for all: Ensuring high-quality STEM education opportunities for all students.* Washington, DC: Author. Retrieved from https://www.whitehouse.gov/sites/default/files/microsites/ostp/stem_fact_sheet_2017_budget_final.pdf

Young, B. A., & Bae, Y. (1997). *Degrees earned by foreign graduate students: Fields of study and plans after graduation* (Issue Brief). Washington, DC: National Center for Education Statistics.

CHAPTER 3

Differentiating Engineering Activities for Use in a Mathematics Setting

Scott A. Chamberlin and Nielsen Pereira

Engineering is truly the very essence of STEM application. This is not to imply that various subdisciplines in mathematics, science, or technology are somehow unimportant, but actually creating products in which knowledge from mathematics, science, and/or technology is utilized enables K–8 students to realize the fruits of their labor. That is to say, mathematics and science are often taught in a vacuum, without context, so actually realizing that they have some use in the real world may be the incentive that problem solvers need to persist in STEM disciplines. Moreover, such applications may enable students not identified as gifted with the use of *typical* or standardized assessments, to surface. Finally, as discussed later, engineering activities may provide opportunities for highly creative products

 DOI: 10.4324/9781003234951-5

to emerge. Therefore, the purpose of this chapter is to discuss the integration of engineering and mathematics and to provide examples of how such integration can be accomplished so that differentiation can be occur.

No concrete evidence, through empirical study, exists to suggest that advanced learners in STEM are any different from advanced learners in other disciplines. Certainly, STEM students' content interests may vary from those who are advanced in other domains, but the needs of such individuals appear consistent with the needs of other advanced students. At the center of their needs is an intense passion to further their understanding of problems and challenging tasks, using differentiation as the vehicle to meet such needs. Consequently, in this chapter, we describe a bridge-building activity and the STEAM Labs program and discuss how the needs of advanced STEM learners may be met, along with the means by which differentiation can be accomplished. An important question pertains to how mathematics and engineering can be integrated with advanced learners. Frankly, the answer to this question varies from classroom to classroom. However, many of the tenets outlined in the Common Core State Standards for Mathematics (CCSSM; National Governors Association Center for Best Practices [NGA] & Council of Chief State School Officers [CSSO], 2010) and the Next Generation Science Standards (NGSS Lead States, 2013) are in harmony with one another and some of these similarities may not be noticeable to stakeholders. For example, Appendix C of the NGSS (NGSS Lead States, 2013) mentions the preparation of students for college and careers, and in the CCSSM, mathematical modeling is the fourth mathematical practice. Incidentally, Lesh, Hoover, Hole, Kelly, and Post (2000) consider mathematical modeling "pre-college level thinking." More specifically, Appendix L of the NGSS, "Connections to the Common Core State Standards-Mathematics," has a detailed table about how the NGSS connect to the CCSSM in grades 2–8 (NGSS Lead States, 2013).

One of the most celebrated features of the NGSS was the emphasis on the engineering design process in addition to the scientific method, which has traditionally been considered the gold standard or a required component of any science class. Many versions of the engineering design process exist and they mostly resemble the scientific method, but with an emphasis on engineering design. An example is the engineering design process used in the Museum of Science, Boston's Engineering is Elementary curriculum, which includes five steps: Ask, Imagine, Plan, Create, and Improve (Hester & Cunningham, 2007). In the first step, students ask questions, such as "What is the problem?" and "What are the constraints?" The Imagine step involves brainstorming possible solutions to a problem and selecting one that will be implemented. The Plan step involves sketching a diagram of the solution and listing the materials to be used, and the Create (often students' favorite) step involves creating, building, and testing a product. Finally, the Improve step allows students to discuss modifications that can lead to a better product (Hester & Cunningham, 2007). Although these steps are presented in this order, which appears to be the most logical way to approach an engineering problem, this version of the engineering design process is presented as a cycle and can start and end at any of the steps.

One may wonder how differentiation can be accomplished in a mathematics setting. In our experience, the best way to accomplish this objective is by presenting to students an engineering problem that allows them to go through the five steps of the engineering design process and includes some form of hands-on activity with a real-life application. Most students enjoy solving problems, but they also appreciate being able to build a product that has some relevance to their lives or culture. The best way to integrate math is by asking students to collect data that allows them to evaluate the quality or effectiveness of their products and then asking them to create some type of mathematical model or to grasp a mathematical concept by using the data they collected. The data can be collected at any of the steps of the engineering design process. For example, students can be provided with constraints or requirements for size or reliability during the Ask step, or those could be introduced as possible ways to Improve their design after students test their product for the first time. Differentiation can be accomplished in many ways, such as the level of support provided throughout the engineering design process or as students create mathematical models; the number of constraints or requirements, which may allow students to be more creative or to simply figure out the solution to a very structured problem; or by the types of questions instructors pose during and after this type of activity.

THE BRIDGE-BUILDING ACTIVITY

This activity is one in which K–8 students design and construct a bridge out of balsa wood. The bridge-building activity (see Table 3.1) has been used for decades, so it is virtually impossible to trace its origins to a single source. Nevertheless, an online search yielded two general templates for the activity (see Resources section of Table 3.1). Teachers are strongly urged to alter the activity to meet their local needs. A general template for this activity is provided at the end of the chapter. Some differentiation will be required for younger students, who may not be able to design and/or construct the bridge without mentoring. One way to aid younger students is by having some pre-made parts for their bridge. Older students, those in grades 6–8, will be able to design and build their bridges with little assistance from an instructor. Although appropriate for K–8 students, this activity will work best with grades 4–8. Another means by which to facilitate dif-

TABLE 3.1
Bridge-Building Lesson Plan

Objectives
Work successfully in teams Design, build, and test a bridge made of balsa wood Collect data on the final bridges once they have been tested Analyze the data and create a mathematical model to reach conclusions and make generalizations

Standards
CCSS.Math.Content.6.SP.A.1. Develop understanding of statistical variability CCSS.Math.Practice.MP4. Model with mathematics NGSS.MS-ETS1. Engineering Design

Materials
Balsa wood, cutting instruments, and wood glue

Instructions
1. Establish expectations and rules for the activity. 2. Identify necessary materials (discuss constraints). 3. Determine a set amount of work time for groups, including an emphasis on the design stage before the build stage. 4. Identify several data points that the entire class should collect (e.g., how much weight was required to break the bridge, what was the height of the bridge at the tallest point, what was the width of the bridge at the widest point, and what was the height of the road [on bridge] from the platform). 5. Discuss why the bridges performed in the manner and make some generalizations about the optimal bridge design.

Variations
o Younger students can be aided by having some pre-made parts for their bridges o Older students (grades 6–8) are able to design and build their bridges with little assistance from an instructor o Hold a contest to see which bridges can hold the most mass before breaking

Resources
Case Western Reserve University. (n.d.). *The Department of Civil Engineering announces a model bridge building contest: Rules*. Retrieved from http://engineering.case.edu/eciv/sites/engineering.case.edu.eciv/files/images/cwru_modelbridge_rules.pdf Garber, D., & Lau, K. (2016). Activity: Balsa wood bridge (ABC-style). *Accelerated Bridge Construction, University Transportation Center, Florida International University*. Retrieved from https://abc-utc.fiu.edu/wp-content/uploads/sites/52/2016/08/3-ABC-Balsa-Wood-Bridge.pdf

ferentiation is to be acquainted with all students' abilities and to challenge them accordingly. For example, a student identified as gifted through an overall battery score may not have the same capabilities to design and build a bridge as one identified as gifted in mathematics or STEM. Hence, the teacher will need to identify a manner in which to aid students less capable in mathematics or engineering or place an additional challenge in front of the student(s) more advanced in STEM.

Implementation Notes

To begin, teachers should establish rules for the activity (e.g., what type of bridge should be built or any requirements/constraints for materials). The dimensions of the bridge should be identified as well. Generally, it is advantageous to have some basic rules so that all bridges have at least several commonalities. For instance, it may be good to demand that all bridges need to have a road/path that has at least 10 centimeters clearance from the testing surface. In addition, instructors may suggest that each bridge must be at least 30 centimeters long and 10 centimeters wide. Furthermore, the manner in which the bridge will be assessed (more on that later) should be identified.

Next, teachers should identify necessary materials. In this activity, it is suggested that at least five pieces of balsa wood (approximately 0.5 cm thickness), with dimensions of 20 centimeters by 20 centimeters be provided. Also, approximately 50 additional sticks of balsa wood, either square or circular in diameter, that are at least 18 centimeters long should be provided. Specifics of the materials are not imperative as long as several considerations are met, and instructors are again strongly encouraged to alter the materials to fit their needs. First, it is good to document the materials used so that they may be easily identifiable for another activity (e.g., next year). Second, problem solvers must have ample materials to create a structurally sound bridge. Third, each group must receive exactly the same materials, without exception. No replacement materials will be provided at any point. In addition to the balsa wood, a cutting instrument, such as a utility knife, and some glue will be necessary. Contingent upon students' level of autonomy, instructors are strongly cautioned to closely supervise knife use or do the cutting for students. In any event, student safety is of utmost importance in this activity given the potentially dangerous equipment. Generally, all of these items can be purchased at a hobby store or online.

Third, teachers should determine a set amount of work time for groups (e.g., 2–3 in-class work days), and then conduct the stress tests. Any type of mass may be used to test the bridges, but two important considerations should be kept in mind. First, problem solvers need to know what the mechanism is to test the bridges prior to designing and building them. Second, it is best if a mass that can be incrementally increased is used. For example, water in a bucket, tied to the center of the bridge, works well because the mass that is added to the bucket can be easily controlled. Please note that a stress test is simply an assessment of what an item can hold before it breaks. Stress tests are conducted on a daily basis on engineering designs, such as airplane wings, ship hulls, bodies of cars, and earthquake-proof structures. A stress test to see how much weight a bridge can hold, therefore, is incredibly realistic and it often is the single best test to determine the best bridge from among several designs.

Instructors may want to hold a contest to see which bridges can hold the most mass before breaking. If a formal contest is not held, pupils will often create their own listing of which bridge performed the best and which performed the worst based on an informal metric. Teachers should be reminded that the stress tests are informative because the ones that may appear strongest may not be the best. A stress test, therefore, will provide accurate data regarding a design's strength. In addition, it is best not to create an excessive number of rules because then pupils may be paying more attention to meeting an extensive set of rules, rather than designing a bridge. Instructors may strongly consider not discussing one type of bridge prior to the activity because the *lecture* may arrest creativity, and problem solvers may build only the type of bridge introduced by the instructor. Alternatively, instructors could host a rather lengthy discussion/analysis of myriad types of bridges prior to the activity in hopes that pupils can select their desired type of bridge. The latter option may provide some differentiation for gifted students. Astute students may not build an aesthetically pleasing bridge (i.e., one they might like to build) when they realize that their bridge will be tested with mass. Finally, instructors are expected to identify several data points that the entire class can collect in an attempt to identify why the bridges performed in the manner in which they did. Instructors must be careful not to give away the answer by only collecting one piece of data. For instance, if an instructor focused solely on the length of the bridge platform, it may encourage pupils to think that designing and building bridges is as simple as having a strong platform. In this activity, all pieces should work symbiotically to create a structurally sound bridge. Examples of various pieces of data that could be collected are weight of each bridge, the amount of mass that it took to destroy the bridge, height, width, length of each bridge, and the type of bridge designed. With data, it is hoped that problem solvers can make some generalizations about the optimal bridge design using balsa wood.

CCSSM and NGSS Meet

As with highly open-ended activities, it is difficult to ensure that all problem solvers will meet specific standards because not all pupils will utilize the same process to solve the problem. That is to say, if seven groups of problem solvers design seven unique solutions, it is problematic to ensure that any one group of standards is met. However, with this activity, it is safe to say that at least two standards, and likely a third one, will be met. First, the domain standard of probability and statistics in the CCSSM (NGA & CCSSO, 2010) will be met because problem solvers will collect data and analyze it in an attempt to make sense of the outcomes, once the bridges have been tested. Second, the general call to implement the engineering design process (NGSS Lead States, 2013) is met because problem solvers will design, build, and then test their bridge with the intent of creating another one at a later date. Third, most problem solvers will meet Mathematical Practice 4 of the CCSSM, model with mathematics, as they analyze the data and quite likely create a mathematical model to reach conclusions and make generalizations. The third standard is not a guarantee and meeting this standard may be contingent upon what pupils have done prior to the bridge activity.

The Beauty of Bridge Building

Some students may be fascinated with the notion that their bridge design is one that is aesthetically pleasing to them. That is to say, some students may not be motivated by the stress test, and they may be motivated to design and construct a particularly creative bridge. It is for this reason that teachers may consider not demanding one type of bridge to be constructed. A lengthy suspension bridge, as an example, may be one such bridge that is incredibly difficult to produce. As a sidenote, one manner in which to assess student affect (for the activity) is to see how many of them design and build another bridge at home or outside of school. Some students may require assistance with this independent work as they may lack the materials and resources to build another bridge. Whenever possible, it is suggested that instructors help problem solvers find additional materials.

The concept of aesthetics and the beauty of a bridge may seem foreign to individuals who restrict the concept of "art" to creative or performing arts. However, many in engineering education consider art an important component of engineering. For instance, some see the Golden Gate Bridge and wonder how long it takes to run across it, and others look at the bridge and realize an amazing artistic component to it. The Golden Gate Bridge was an engineering marvel, as nothing like it had ever been constructed. It could be hypothesized that the engineers designed and built it as much for convenience of citizens in the Bay Area inasmuch as they designed and constructed it just to see if they could do it. Amazingly, the Golden Gate Bridge is now 80 years old and was the longest suspension bridge in the world for 27 years. Moreover, the bridge-building activity may be a positive introduction to the STEAM Labs program (Jordan, Dalrymple, Pereira, Astatke, & Fletcher, 2012; Jordan & Pereira, 2008, 2009), described in the next section.

STEAM LABS

The STEAM Labs program challenges middle and high school students to apply the engineering design process in creative ways. In addition to studying topics in science, technology, engineering, and math, students also focus on the arts (the "A" in STEAM). Through the STEAM Lab programs, students design and build STEAM Machines™, which are chain reaction machines that complete simple tasks in complex and roundabout ways. These machines resemble the ones portrayed in Rube Goldberg cartoons and they include a chain reaction of steps (i.e., the smallest action that is performed by a machine), which comprise the various modules in a machine. STEAM Machines™ usually involve a common theme (i.e., to create some unity among the various modules), storyboarding, and background music related to the theme or storyboarding. Students work in groups, which are initially responsible for designing and building modules, and then the modules are integrated into a whole-class machine or even connected to machines at other sites (sometimes in a different country).

Goals for the STEAM Labs program include developing student creativity, and, unlike many STEM classes and programs, students are encouraged to be as creative as possible while designing and building their STEAM Machines™. This means there is no limit to how complex or unique the machines will be. Students are challenged with learning and applying the Engineering is Elementary (Hester & Cunningham, 2007) engineering design process, which was developed by the Museum of Science, Boston. Students are also expected to participate in multiple brainstorming sessions while they design, build, test, and think of ways to improve their machines. One math concept included in the STEAM Labs curriculum is reliability, as students are encouraged to create STEAM Machines™ that are as reliable as possible. As an introduction to the idea of chain-reaction machines, instructors show videos of machines designed and built by Rube Goldberg enthusiasts and other students. One great source of such videos is the Rube Goldberg website (https://www.rubegoldberg.com), which includes an archive of photos and videos of the Rube Goldberg Machine Contest participants. In order to introduce the engineering design process, students brainstorm and design their first STEAM Machine™ as a class, and then each team in a classroom (the ideal number of team members is 3–4 students) builds its own version of that design. The results are usually that each team builds a machine that includes all of the steps and components in the design created by the class, but machines may look completely different from one another as teams might have different approaches to implementing that design, such as using components of different lengths and sizes. That realization of the differences and similarities among the products created by students in itself is an important lesson for students involved in the STEAM Labs™ program.

Needless to say that several of the implementation notes, differentiation strategies, application of math concepts, and standards described in the previous sections also apply to the STEAM Labs program. Follow-up activities include holding a contest to see which machine is the most reliable, includes the most/least steps, or involves the most creative uses of the available materials; or asking groups to integrate their modules into a whole-class machine, which often requires some redesign, such as adding or removing steps to connect modules. This extension activity provides an opportunity to further discuss reliability and the effect of each individual module on overall reliability (e.g., what happens if a module that is 100% reliable is connected to one that only works 50% of the time?). The opportunities for discussing math and engineering applications are endless with this type of activity, and there is no ceiling as to how creative students can be while designing and building STEAM Machines™. Standards addressed in the STEAM Labs program include engineering design (NGSS.MS-ETS1; NGSS Lead States, 2013) and the CCSSM (NGA & CCSSO, 2010) related to probability models (e.g., 7.SP.C.5 and 7.SP.C.6).

TEACHER CONTENT DISCUSSION

Instructors are expected to have two to three content talking points to conclude activities integrating math and engineering and any engineering inquiry activity. In short, educators are no longer in a position to simply provide cute or fun activities to students. Moreover, engaging activities can be bolstered considerably with a robust educational discussion. The term *discussion* is used in this instance to suggest that there is some give and take between the instructor(s) and the problem solvers. A discussion, therefore, occurs when an instructor provides talking points and uses open-ended questions to facilitate student thinking. To that end, instructors should have a series of questions, closely linked with their content objectives, for use in a discussion. A basic set of questions can be modified as needed to investigate follow-up points. A set of questions for the bridge-building activity may look something like this:

- How did your design and ultimate construction help or hurt the performance of your bridge?
- Did the length the bridge have anything to do with its structural integrity (in engineering, an object's ability to withstand high amounts of stress is referred to as its structural integrity)?

- Using the data that we collected on the bridges (e.g., weight, strength), can you develop an explanation as to why each bridge broke as it did?
- How might your bridge have held up better if you had some other type of material to construct it? (This is a great question for differentiation.)

CONCLUSION

The bridge-building activity and the STEAM Labs programs are examples of how engineering and mathematics can be integrated seamlessly. Problem solvers engage in the design, construction, and ultimate demise of their product. However, the authentic nature of these activities is one that is likely to keep problem solvers engaged, perhaps long after the activity concludes. These activities are linked with at least two, and likely three, standards from the CCSSM (NGA & CCSSO, 2010) and the NGSS (NGSS Lead States, 2013) and are inherently set up for differentiation, pending the needs of each classroom. Perhaps the best recommendation for differentiation pertains to how products could be constructed again using the conclusions reached with the data. Alternatively, the use of different materials might precipitate considerably different outcomes and designs among problem solvers.

REFERENCES

Hester, K., & Cunningham, C. M. (2007). *Engineering is elementary: An engineering and technology curriculum for children.* Paper presented at the American Society for Engineering Education Annual Conference and Exposition, Honolulu, HI. Retrieved from http://eie.org/sites/default/files/research_article/research_file/ac2007full8.pdf

Jordan, S. S., Dalrymple, O., Pereira, N., Astatke, Y., & Fletcher, J. D. (2012). *Design swapping as a method to improve design documentation.* Paper presented at the American Society for Engineering Education Annual Conference and Exposition, San Antonio, TX.

Jordan, S., & Pereira, N. (2008). *Design twice, build once: Teaching engineering design in the classroom.* Paper presented at the National Center for Engineering and Technology Education Conference on Research in Engineering and Technology Education, St. Paul, MN.

Jordan, S., & Pereira, N. (2009). *Rube Goldbergineering: Lessons in teaching engineering design to future engineers.* Paper presented at the American Society for Engineering Education Annual Conference and Exposition, Austin, TX.

Lesh, R., Hoover, M., Hole, B., Kelly, A., & Post, T. (2000). Principles for developing thought-revealing activities for students and teachers. In A. Kelly & R. Lesh (Eds.), *The handbook of research design in mathematics and science education* (pp. 591–646). Hillsdale, NJ: Lawrence Erlbaum.

National Governors Association Center for Best Practices, & Council of Chief State School Officers. (2010). *Common Core State Standards for mathematics.* Washington, DC: Author.

NGSS Lead States. (2013). *Next Generation Science Standards: For states, by states.* Washington, DC: The National Academies Press.

CHAPTER 4

Creating a Climate of Inventiveness, Innovation, and Creativity

Laurie J. Croft

Stories about amazing new ideas or products fuel our collective imaginations. *Time for Kids*, for example, publishes a popular annual review of the year's new inventions, and its editors talk about their selections on YouTube (e.g., https://www.youtube.com/watch?v=yTRkdm-hi04). The inventions span from gadgets, robots, and technology, to clothing, medicine, miscellaneous, and more. Other magazines (e.g., *Consumer Reports, Popular Mechanics, Popular Science*) offer articles or annual editions dedicated to new concepts that meet our needs or match our interests in better ways. When the ideas originate with "young innovators who've already made advances in the fields of science and technology before they're even old enough to vote" (Siegel, 2011, para. 1), we want to encourage, and to capitalize on, the imaginations, creativity, and inventive frames of mind of the young people in our communities.

DOI: 10.4324/9781003234951-6

INVENTION AND INNOVATION AS CURRICULUM

The story of our civilization is based on invention; therefore encouraging students to invent is important for their future and, from a global perspective, our future. (Westberg, 1996, p. 265)

A dynamic process, inventiveness combines the practical application of knowledge and theory with the creative impulse to make something better, and invention and innovation programs align neatly with Dimension 1: Science and Engineering Practices of the Next Generation Science Standards (NGSS; Adams, Cotabish, & Ricci, 2014). Even though more patents are held by men (Basken, 2006), the K–12 invention process supports the creative endeavors of any student, regardless of gender, race, or socioeconomic status. Adult inventors report generally liking school but not excelling in academics (Colangelo, Assouline, Kerr, Huesman, & Johnson, 1993), and in school-based inventiveness programs, young inventors are often those who are taking their toys apart rather than playing with them and tinkering with anything else they can disassemble. The inventive process often appeals to students characterized by practical intelligence (Sternberg, Ferrari, Clinkenbeard, & Grigorenko, 1996), a form of intellectual ability often overlooked in traditional classrooms. Invention programs explain the patent process, often include lessons exploring entrepreneurship, and encourage students to find a problem to solve in the world around them. Inventors, often committed to making things better, simpler, and generally wanting to contribute to society (Colangelo et al., 1993), have urged educators to teach more about innovation (e.g., British inventor Trevor Baylis: "Kids today need to put down their mobile phones and start tinkering" [Burn-Callander, 2013]).

Inventive thinking programs gained popularity in American schools in the 1980s, endorsed by leaders in business, science, and technology, and promoted by agencies as prestigious as the United States Patent and Trademark Office (USPTO). Different programs provided outreach, including free resources, to educators who wanted to encourage both creative thinking and problem solving. Invent Iowa, for example, housed since 1989 at the Connie Belin and Jacqueline N. Blank International Center for Gifted Education and Talent Development at the University of Iowa, has challenged students to invent. The process is multidisciplinary and develops or strengthens academic skills, including reading, research in both the library and real-world settings, divergent and convergent thinking skills, utilization of scientific and technological concepts, writing, design, and persuasive thinking. Programs such as Invent Iowa have provided curriculum and programming support, and they often culminate in invention conventions, allowing students to showcase their work and to explain their thinking, their process, and their product to an authentic audience.

AN INVENTIVE PROCESS FOR STUDENTS

To understand invention and discovery, students need to become creative producers by working actively and collaboratively on open-ended, real-world problems to which there is no single right answer. (Gorman, Plucker, & Callahan, 1998, p. 530)

The Invent Iowa process exemplifies the variety of options that have emerged since the end of the 20th century and are now available online for educators, parents, or students who search for "invention curriculum." The organizational features include an introduction to inventing that features an overview of inventing, including consideration of the differences between "invention" and "innovation," and an exploration of the significant contributions inventors have made to society. Invention education is, by nature, fun for students, especially when the process includes an opportunity to create Rube Goldberg[2] devices with random materials made available for the activity. Westberg (1996) found that basic introductory lessons about inventing can increase the number of student inventions (Westberg, 1996), but these introductions do not improve the quality of student work. Plucker (2002) suggested that a single introductory lesson fails to help students understand that inventors spend a great deal of time on their work, reflecting on and tinkering with their inventions so that others will recognize the value of their ideas.

In order to ensure that "creative scientific skills are teachable, useful for students, and desirable as outcomes of science education" (Plucker, 2002, p. 150), the general invention process needs to continue with steps that are similar to many of the early steps utilized by those who identify themselves as inventors:

- brainstorming ideas or utilizing other activities to identify needs or wants;
- recording ideas and their development in a notebook or log, witnessed by others;
- considering possible solutions, and studying the content areas that inform and develop those solutions;
- designing prototypes or models;

---

2   Named for Rueben Goldberg (1983–1970), a cartoonist, engineer, and inventor of devices that provided humorously convoluted solutions for essentially simple tasks.

- reviewing strategies for making a new product both accessible and desirable to those who would benefit most from its availability, including naming the product; and
- showcasing inventions in settings that allow young inventors to receive practical and effective feedback/evaluation (Gorman & Plucker, 2003; Plucker & Gorman, 1995; Shlesinger, Jr., 1980; Westberg, 1996).

Although invention conventions would not be a part of the professional inventor's experience prior to the patent process, for many students, the possibility of participating in a competition designed for young inventors enhances a sense of autonomy and the importance of self-directed learning skills, process skills, and both personal and interpersonal skills (Karnes & Riley, 1999). Young inventors begin to appreciate the invention process as an active rather than static one, internalizing that understanding through active participation. Open-ended problems encourage students to start "taking the sorts of risks an inventor would take" (Gorman et al., 1998, p. 531), and to appreciate peers who could contribute to challenges that demand competencies ranging from imagination, creativity, spatial ability, mechanical aptitude, scientific or technical knowledge, artistic ability, and entrepreneurship.

Introducing students to one or more problem-solving models is central to inventive thinking learning modules. Invent Iowa describes steps that inventors could use after identifying a problem that they want to solve, including the careful consideration of the "5 W's and 1H" (Baldus, Blando, & Croft, 2005, p. 98); that is, inventors ask "who, what, when, where, why, and how?" when thinking about their ideas. Inventors continue revisiting these questions as they imagine their design and begin creating it. Invent Iowa introduces various problem-solving models, including, for example, five basic steps following the identification of a problem: (1) determining relevant facts, (2) considering underlying problems, (3) developing applicable ideas, (4) finding potential solutions, and (5) developing a plan of action. Following the recognition of a problem, an inventor begins to more clearly define the problem, devise possible solutions, evaluate each possibility in order to select the best one, implement a plan to build the invention, and evaluate the results. This type of modeling for the novice inventor encourages students to go through cycles of planning, monitoring their progress, and reflecting on how effective their inventions will be, as well as how to improve them (Beyer, 1987). The process is very similar to the cycle of design that engineers use when solving problems (Sneider, 2012); problem-solving models for inventive thinking develop habits of mind that facilitate student understanding and use of engineering design.

21ST-CENTURY NEED FOR INVENTIVE THINKING

> Innovation means that children are going to have to have developed at some point in their educational careers a love for learning, a passion for learning. This allows them to be adaptable and successful and a part of the competitive workforce permanently. (Utah Governor Jon Huntsman, Jr., as cited in Fitzpatrick, 2007, p. 23)

In the first decade of the 21st century, invention programs have shifted to the periphery of the educational environment, largely as extracurricular options. School districts have been faced with high-stakes assessments, and they have wanted teachers to present lessons that prepare students for tests. Teachers have been provided with scripted (sometimes even paced) curricula, and they have had few opportunities to encourage innovative thinking. Referencing the importance of vital educational programs in STEM, a National Research Council (NRC; 2011) report noted that "current assessments limit teachers' ability to teach in ways that are known to promote learning of scientific and mathematical content practices" (p. 27), replacing complex performance assessments with multiple-choice items. Darling-Hammond (2000) found an "almost inverse relationship between statewide testing policies and both teaching standards and student performance" (p. 23), perhaps because many teachers narrowed their curriculum to more closely match carefully defined content expectations.

Contemporary society, however, is seeking to focus on advanced skills: "The need for knowledge workers, to innovate and create new products and services that solve real problems, is a major force driving the world economy" (National Governor's Association, 2012, p. 8). Disciplinary standards such as the NGSS provide clues for reaching both short- and long-term goals, encouraging lessons that promote real-world and open-ended problems that reflect 21st-century needs. The NGSS provide opportunities for teachers to build on the three dimensions that encourage deep understanding of scientific reasoning, including Science and Engineering Practices, Crosscutting Concepts that link different domains in science, and Disciplinary Core Ideas that emphasize the most important concepts in the field. Educators can differentiate for the needs of gifted and talented learners by emphasizing real-world applications of students' developing academic understandings, thus overcoming "a variety of challenges that stand between vision and reality" (NRC, 2011, p. 19). Necessary teacher expertise includes appropriate preparation in STEM content, as well as in the nature and needs of students who have advanced abilities and skills in the field of study. Teachers can be most effective by practicing careful and purposeful lesson planning, delivering well-crafted curriculum with appropriate pacing, using higher order questions and wait time for the development of student answers, and embedding assessments that measure student understanding throughout the learning process (Tweed, 2009; Varella, 2000).

A LESSON PLAN FOR INVENTIVE THINKING

A sample lesson plan has been adapted from the *Invent Iowa Curriculum Guide* (Baldus et al., 2005, pp. 57–58) and provided in a backward design format (Wiggins & McTighe, 2005). The lesson plan, detailed in Table 4.1, challenges students to design a container to preserve an egg dropped from 8 feet and emphasizes the Engineering, Technology, and Applications of Science standards of the NGSS. Additional science concepts could be introduced in conjunction with the activity. The Egg Drop project (https://stem.neu.edu/resources/activities/eggdrop), for example, facilitates student learning about momentum, pressure, air resistance, angular momentum, and gravity. The National Aeronautics and Space Administration (NASA; 2012) connects its "Mars Pathfinder Egg Drop Challenge" with the need for the soft landing of a spacecraft on Mars.

Central to the *Invent Iowa Curriculum Guide* is Unit 10, Student Inventing Practice, in the intermediate invention activities (Baldus et al., 2005; see Table 4.1). The lesson builds on multiple understandings and skills reflected in the primary (grades K–2) units, as well as in earlier intermediate (grades 3–5) units. These include general understandings of why humans invent, historical perspectives on inventors and their inventions, the diversity among inventors (and obstacles that women and minorities have faced in obtaining patents and recognition), relevant vocabulary (e.g., patents, invention, innovation), and possible steps for inventing (e.g., Shlesinger, Jr., one of many options provided in Baldus et al., 2005). The lesson can be implemented in a variety of settings, from general education or gifted pullout classrooms, to enrichment or extracurricular programs. Following this opportunity to innovate a solution to a problem provided by the teacher, students can continue their Inventing Practice by identifying a new problem to solve—one that can make a difference in their lives, and in the lives of their families, friends, or communities.

TABLE 4.1
Backward Design Lesson Plan: Student Inventing Practice

Stage 1: Desired Results		
Established Goals	**Transfer**	
3-5-ETS1 Engineering Design. Students who demonstrate understanding can:	*Students will be able to independently use their learning to . . . invent something that will pre-vent an egg from breaking when dropped from an 8-foot ladder.*	
o 3-5-ETS1-1. Define a simple design problem reflecting a need or a want that includes specified criteria for success and constraints on materials, time, or cost.	**Meaning**	
	Understandings	**Essential Questions**
	Students will understand that . . .	1. How do effective problem solvers develop new solutions to problems?
o 3-5-ETS1-2. Generate and compare multiple possible solutions to a prob-lem based on how well each is likely to meet the criteria and constraints of the problem.	o People's needs and wants change over time, as do their demands for new and improved technologies (3-5-ETS1-1).	2. How do inventors deal with challenges to the design process?
	o Engineers improve existing technologies or develop new ones to increase their benefits, decrease known risks, and meet societal demands (3-5-ETSI-2).	3. How do engineers deal with challenges to the design process?
o 3-5-ETS1-3. Plan and carry out fair tests in which variables are controlled and failure points are considered to identify aspects of a model or proto-type that can be improved.	**Acquisition**	
	Students will know . . .	*Students will be skilled at . . .*
	o possible solutions to a problem are limited by available materials and resources (constraints).	o utilizing a problem-solving methodology to explore the development of possible inventions/innovations, but understanding that the steps are not inflexible and can be implemented in different ways.
	o the success of a designed solution is deter-mined by considering the desired features of a solution (criteria).	o maintaining an inventor's journal to

KEY COMPONENTS OF ENGINEERING INSTRUCTION

TABLE 4.1, *CONTINUED*

o different proposals for solutions can be compared on the basis of how well each one meets the specified criteria for success or how well each takes the constraints into account (3-5-ETS1-1-A; ETS1.A: Defining and Delimiting Engineering Problems). o research on a problem should be carried out before beginning to design a solution. o testing a solution involves investigating how well it performs under a range of likely conditions (3-5-ETS1-2). o that communicating with peers about proposed solutions is an important part of the design process, and shared ideas can lead to improved designs (3-5-ETS1-2). o tests are often designed to identify failure points or difficulties, which suggest the elements of the design that need to be improved (3-5-ETS1-3; ETS1.B: Developing Possible Solutions). o different solutions need to be tested in order to determine which of them best solves the problem, given the criteria and the constraints (3-5-ETS1-3; ETS1.C: Optimizing the Design Solution).	o capture the development process, from sketches of initial ideas to data acquired to support possible solutions. o receiving feedback from peers about possible solutions, and providing feedback to peers about their ideas. o apply decision-making skills to determine best possible solution, considering necessary materials and costs. o building a model of an invention to protect an egg.

TABLE 4.1, *CONTINUED*

Stage 2: Evidence	
Evaluative Criteria	**Assessment Evidence**
Inventor's journal reflecting all steps of inventing: o Identification of problem/need o Foundation of current understanding (and brainstorming for possible solutions) o Data related to possible solutions, including knowledge and skills needed for solution o Imagination in applying decision-making skills o Limitations in proposed invention/innovation and modifications o Test of container	**Performance Task(s)** *Packaging challenge: What type of container will prevent an egg from breaking when dropped from an eight-foot ladder?* This exercise builds on multiple understandings and skills reflected in the primary (grades K–2) units, as well as in earlier intermediate units. These include general understandings of why humans invent, historical perspectives on inventors and their inventions, the diversity among inventors (and obstacles that women and minorities have faced in obtaining patents and recognition), relevant vocabulary (e.g., patents, invention, innovation), and possible steps for inventing (e.g., Shlesinger, Jr.).
Collaboration with peers/mentors	**Other Evidence** o Courteously receive feedback from others about possible solutions (understand this is essential in engineering, but sometimes not practiced among inventors, at least in initial stages) o Courteously provide useful feedback to peers about possible solutions during class debriefing o Consider design shortcomings and revisit invention process to improve design

TABLE 4.1, *CONTINUED*

Stage 3: Learning Plan

1. Review Shlesinger, Jr's steps for inventing.
  a. **Identification:** Individually brainstorm problems related to eggs; record ideas in your Inventor's Journal and share ideas in small groups.
  b. **Foundation:** What do you already know about packing materials and cartons? In your group, discuss 3–5 possible solutions for a container that will prevent an egg from breaking when dropped from 8 feet.
  c. **Data:** In your small group, review the "5 W's and 1 H" question process; try to think of questions about this scenario that help you define necessary information for your invention. Add to your Inventor's Journal, chart information that helps you, and seek any knowledge and/or skills that support your efforts.
  d. **Imagination:** Use decision-making skills to decide on the solution you want to try. Sketch ideas in your Inventor's Journal and label them. List materials you need, and possible costs for the invention. Be prepared to revise your ideas as you learn new information.
  e. **Limitations:** As a whole group, share ideas and sketches, and consider potential problems. Using available materials, build a model of your invention. Implement your solution, and alter your design if necessary.
2. Review the steps for inventing in the context of the egg drop.

Note. Adapted from *Invent Iowa Curriculum Guide* (pp. 57–60), by C. Baldus, C. Blando, and L. Croft, 2005, Iowa City: Universty of Iowa, The Connie Belin & Jacqueline N. Blank International Center for Gifted Education and Talent Development. Copyright 2005 by The Connie Belin & Jacqueline N. Blank International Center for Gifted Education and Talent Development. Adapted with permission.

RESOURCES

Although some of these resources appeal directly to students, others provide recommendations for teachers and/or parents. Some provide access to materials available at no charge, others have links to pages with materials for sale and others provide information about afterschool programs and/or invention camps.

- **BKFK Education (http://bkfkeducation.com):** BKFK describes itself as an agency dedicated to supporting the innovative spirit within young innovators, as well as providing access to marketing and media relations. The site has links for parents, educators, and students.
- **Davidson Institute for Talent Development (http://www.davidson gifted.org/db/browse_resources_212.aspx):** The Davidson Institute provides services and programs to recognize and support profoundly intelligent young people in the areas that will benefit them individually. Young people apply each year to become Davidson Fellows, recognized for their talent in areas as diverse as mathematics, science, literature, music, technology, philosophy and "outside the box." A variety of links provide access to competitions and information for young inventors.
- **Invent Iowa (http://www2.education.uiowa.edu/belinblank/Students/ inventia):** Invent Iowa serves the needs of talented young inventors and their teachers—participants are welcome from anywhere in the world. The Invent Iowa program encourages students to creatively think and solve problems through the invention process. The site includes freely available curricula and other resources.
- **Invent Now, Inc. (http://www.inventnow.org):** Invent Now has multiple pages to interest student inventors and provides an opportunity for students to register as an inventor (no cost; no last name, address, or other identifying information). Registered inventors have access to a newsletter, and games, and can go through a simulation of applying for a patent. InventNow sponsors Camp Invention (http://campinvention. org).
- **"Inventions & Science History" by The Science Spot's Kid Zone (http://sciencespot.net/Pages/kdzinvent.html):** Multiple links provide information of interest to young inventors, from Ancient Inventions and Forgotten Inventions to the Inventor of the Week and Explorit Science Center Daily Facts. The site also provides links to invention and patent companies.
- **Kids Invent! (http://www.kidsinvent.org):** In addition to limited free information for children ages 7–15, creative learning activity kits are available for sale. Links provide information about Kids Invent! camps

and afterschool programs, science fair projects, and science standards linking the creative learning activity kits to specific standards.

- **Lemelson Center for the Study of Invention and Innovation (http://invention.si.edu):** The Lemelson Center encourages students and adults to better understand technological, economic, and social change. Embedded in the Smithsonian Institute, the Center undertakes historical research, develops educational initiatives, creates exhibitions, and shares invention stories that inform and inspire.
- **USPTO Kids (http://www.uspto.gov/kids):** The United States Patent and Trademark Office advises the president—and provides educational resources and support for children's passion and creativity. Believing that all children have the potential to do extraordinary things, they encourage innovation as a tool for life-long learning.

REFERENCES

Adams, C. M., Cotabish, A., & Ricci, M. C. (2014). *Using the Next Generation Science Standards with gifted and advanced learners.* Waco, TX: Prufrock Press.

Baldus, C., Blando, C., & Croft, L. (2005). *Invent Iowa curriculum guide.* Iowa City: University of Iowa, the Connie Belin & Jacqueline N. Blank International Center for Gifted Education and Talent Development. Retrieved from http://www2.education.uiowa.edu/belinblank/Students/inventia/docs/inventia.pdf

Basken, P. (2006, August 4). Women scientists lag far behind men in patents, study says. *The Boston Globe.* Retrieved from http://www.boston.com/news/nation/articles/2006/08/04/women_scientists_lag_far_behind_men_in_patents_study_says

Beyer, B. K. (1987). *Practical strategies for the teaching of thinking.* Boston, MA: Allyn & Bacon.

Burn-Callander, R. (2013, September 3). British inventor: Trevor Baylis calls for schools to teach importance of invention. *The Telegraph.* Retrieved from http://www.telegraph.co.uk/finance/newsbysector/mediatechnologyandtelecoms/10281540/British-inventor-Trevor-Baylis-calls-for-schools-to-teach-importance-of-invention.html

Colangelo, N, Assouline, S. G., Kerr, B. A., Huesman, R., & Johnson, D. (1993). Mechanical inventiveness: A three-phase study. In G. R. Bock & K. A. Krill (Eds.), *The origins and development of high ability* (pp. 160–174). Chichester, England: John Wiley.

Darling-Hammond, L. (2000). Teacher quality and student achievement. *Education Policy Analysis Archives, 8*(1), 1–44. http://dx.doi.org/10.14507/epaa.v8n1.2000

Fitzpatrick, E. (2007). *Innovation America: A final report.* Washington, DC: National Governors Association. Retrieved from http://www.nga.org/files/live/sites/NGA/files/pdf/0707INNOVATIONFINAL.PDF

Gorman, M., & Plucker, J. (2003). Teaching invention as critical creative processes: A course on technoscientific creativity. In M. A. Runco (Ed.), *Critical creative processes* (pp. 275–302). Cresskill, NJ: Hampton Press.

Gorman, M. E., Plucker, J. A., & Callahan, C. M. (1998). Turning students into inventors: Active learning modules for secondary students. *Phi Delta Kappan, 79,* 530–535.

Karnes, F., & Riley, T. (1999). Developing an early passion for science through competitions. *Gifted Child Today, 22*(3), 34–36.

National Aeronautics and Space Administration. (2012). *Lesson title: Mars Pathfinder egg drop challenge.* Retrieved from https://www.nasa.gov/offices/education/programs/national/summer/education_resources/engineering_grades7-9/E_egg-drop.html#.V6zlCbW4jnQ

National Governors Association. (2012). *New engines of growth: Five roles for arts, culture, and design.* Washington, DC: Author.

National Research Council. (2011). *Successful K–12 STEM education: Identifying effective approaches in science, technology, engineering, and mathematics.* Washington, DC: Author.

Plucker, J. (2002). What's in a name? Young adolescents' implicit conceptions of invention. *Science Education, 86,* 149–160.

Plucker, J., & Gorman, M. E. (1995). Group interaction during a summer course on invention and design for high ability secondary students. *The Journal of Secondary Gifted Education, 6,* 258–272.

Shlesinger, Jr., B. E. (1980). I teach children to be inventors. *Educational Leadership, 37,* 572–573.

Siegel, R. (2011, December 28). Young innovators: Modifying Microsoft's Xbox Kinect. *NPR.* Retrieved from http://www.npr.org/2011/12/27/144335732/teens-win-top-honors-for-xbox-innovation

Sneider, C. (2012). *Core ideas of engineering and technology: Understanding a framework for K–12 science education.* Retrieved from http://www.nsta.org/docs/ngss/201201_Framework-Sneider.pdf

Sternberg, R. J., Ferrari, M., Clinkenbeard, P. R., & Grigorenko, E. L. (1996). Identification, instruction, and assessment of gifted children: A construct validation of a Triarchic Model. *Gifted Child Quarterly, 40,* 129–137.

Tweed, A. (2009). *Designing effective science instruction: What works in science classrooms.* Arlington, VA: National Science Teachers Association.

Varella, G. (2000). Science teachers at the top of their game: What is teacher expertise. *Clearing House, 74,* 43–46.

Westberg, K. L. (1996). The effects of teaching students how to invent. *The Journal of Creative Behavior, 30,* 249–267.

Wiggins, G., & McTighe, J. (2005). *Understanding by design* (2nd ed.). Alexandria, VA: Association for Supervision and Curriculum Development.

CHAPTER 5

Enhancing K–8 Engineering Through Arts Integration

Rachelle Miller and Callie Slider

Engineers have contributed to our society for hundreds of years. For instance, civil engineers have produced architectural wonders such as the Great Wall of China, the Golden Gate Bridge, the pyramids of Egypt, and many more. Mechanical engineers influence the design of luxury automobiles like BMW, Lexus, and Porsche. What attracts us to these wonders? Is it their functional designs, the aesthetic components, or both? Initially, the arts and sciences appear to be two distinct and unrelated domains. However, the relationship between art and science began hundreds of years ago; the Renaissance period can be described as the peak of its integration. By the late 1800s and early 1900s, major paradigm shifts occurred in art and science and these two areas became more distinct. Specialization of fields became more common and universities began to offer separate programs for art and science, thus making it difficult to pursue both fields.

    DOI: 10.4324/9781003234951-7

In recent years, more attention has been given to the relationship of these two domains and the reciprocal benefits that they can offer.

OVERVIEW

This chapter consists of four sections: parallels between the arts and engineering, integrating the arts and creativity with engineering, arts integrated lessons, and suggested strategies for arts integration within other content areas.

PARALLELS BETWEEN ENGINEERING AND THE ARTS

There are similarities between the problem-solving procedure that engineers use and the creative process that artists experience (Robinson & Baxter, 2013). Both identify a problem, research the problem, and explore ways to solve the problem. Engineers focus on designing or improving a product to solve the problem and artists focus on creating an art form with the most appropriate medium. The commonalities are illustrated in Table 5.1.

Various individuals throughout history exhibited both artistic abilities and engineering skills that led to the development of innovative products. The most prominent historical figure is Leonardo da Vinci, who is known as a talented artist and who was also a gifted engineer. Da Vinci is well known for creating the *Mona Lisa* and *The Last Supper*, and he was also well ahead of his time in the discovery and design of the following advanced engineering concepts and skills: line of thrust used to develop arches (civil engineering), design and experimentation of flying machines (aeronautic engineering), design of military equipment (military engineering), and the interactions between different machines (systems engineering; The Engineer, 2006; Sniderman, 2012). Another example is Samuel F. B. Morse, an art professor, who invented the first telegraph system in the United States. Morse's career began as an artist and then later as a communications engineer. Several other researchers have also indicated relationships among engineering, innovative thinking, and the arts (Alias, Black, & Grey, 2002; LaMore et al., 2013; Root-Bernstein et al., 2008).

TABLE 5.1
*Similarities Between Engineering and
Artistic Problem Solving Processes*

Engineers	Artists
Define the problem	Select concept to portray
Collect known information	Examine existing art forms
Determine what is unknown	Identify challenges to portraying the concept
Gather appropriate tools	Gather appropriate materials
Test the proposed product	Select appropriate medium for art form

INTEGRATING THE ARTS AND CREATIVITY WITH ENGINEERING

Teaching art through an engineering context enhances problem-solving skills, critical thinking skills, creativity, and innovative thinking (Orhun & Orhun, 2013; Robinson & Baxter, 2013). Oftentimes, engineering courses focus heavily on the procedural approach to problem solving and less on students evaluating and using creativity to synthesize the problem (Vuksanovich & Wallace, 2011). According to the National Academy of Engineering (2004),

We aspire to an engineering profession that will rapidly embrace the potentialities offered by creativity, invention, and cross-disciplinary fertilization to create and accommodate new fields of endeavor, including those that require openness to interdisciplinary efforts with nonengineering disciplines. (p. 50)

ARTS INTEGRATED LESSON PLANS FOR GIFTED STUDENTS

Arts integration is a teaching strategy that uses interdisciplinary curriculum to connect the arts with another content domain. This instructional approach includes lesson objectives for both the arts and the other content domain and results in a deeper understanding of the content (Silverstein & Layne, 2010). This type of instruction supports the differentiation of gifted students by encouraging the teacher to use multiple means of content representation and providing students the opportunity to express their knowledge of the content in various ways (Lynch, 2007).

During Summer 2015, the University of Central Arkansas offered a 4-day STEMulate Academy, the state's first engineering summer academy for gifted and talented students in grades 3–5. Engineering is Elementary (EiE; Museum of Science, Boston, 2016) was the primary resource used in each grade level, and the content, process, and products were differentiated for gifted students. In addition, an arts specialist developed arts integrated lessons that incorporated the arts with engineering content. Prior to the academy, teachers completed a one-day professional development session on their *EiE* unit. They also worked side-by-side with an art teacher to help them effectively integrate the arts into their aerospace, electrical, and transportation engineering units for gifted students.

Grade 3 students completed the engineering unit, *A Long Way Down: Designing Parachutes*. Color study was integrated in this aerospace engineering unit by using a chromatography activity (Knighten, 2016) to decorate their parachutes. Throughout the chromatography lesson, students explored, examined, and discovered the hidden primary colors that make up water-soluble markers. Furthermore, they investigated the differences between paper and gas chromatography and how gas chromatography is used in aerospace engineering.

Visual arts, art history, graphic arts, and literary arts were integrated into the fourth-grade unit, *The Attraction Is Obvious: Designing Maglev Systems*. Students combined their knowledge of magnets and connectivity with visual art to create a Jackson Pollock-inspired painting (Left Brain Craft Brain, 2015). The students continued the project by creating advertisements for their new maglevs to convince people to buy their product and by writing letters to someone who lived more than 100 years ago to describe the maglev system and its importance. (See Table 5.2 for lesson details on incorporating the arts into this unit.)

Two additional units that integrated art and engineering were *An Alarming Idea: Designing Alarm Circuits* and *Slick Solution: Cleaning an Oil Spill*. Visual arts, art history, and literary arts were integrated into the fifth-grade electrical engineering unit titled *An Alarming Idea: Designing Alarm Circuits*. Students designed original constellations (Meethumalu, n.d.) using circuits and created storyboards to write about their knowledge of circuits. (See Table 5.3 for lesson details on integrating the visual arts and art history into this unit.) Visual arts can also be incorporated into the EiE unit *A Slick Solution: Cleaning an Oil Spill*. Students can learn how environmental engineering practices are relevant to our lives through the lens of arts. (See Table 5.4 for arts integration suggestions.)

TABLE 5.2
Suggestions for Integrating Art and Maglev Systems

Pulling Paint
Students will combine their knowledge of magnets and connectivity with visual art to create a Jackson Pollock-inspired painting.

Materials

o   Assorted small metal objects (e.g., paperclips, bolts, screws, washers)
o   Strong magnets
o   Paper plates
o   Paint (acrylic or tempera)
o   Educational materials on Jackson Pollock

Suggestions: Action Jackson by Jan Greenberg is a great book for kids. "Jackson Pollock 51" is also a great resource on YouTube: https://www.youtube.com/watch?v=CrVE-WQBcYQ.

Procedure

o   Students will choose a set quantity of paints. *(Suggestion: Four or five is plenty.)*
o   Students will add quarter-size dollops of paint on their plate. *(Suggestion: Disperse them evenly on the plate. You don't want them to touch or mix prior to the activity.)*
o   Carefully place the metal objects ANYWHERE on the plate.
o   Have a partner hold the plate while another student takes the magnet and moves it around underneath the plate. *(You should see the magnets grabbing the metal objects and dragging them through the paint.)*
o   Have students write a parallel paragraph that describes how the process of making this painting parallels a maglev system.

Design Your Own Maglev Transportation
Students will combine their knowledge of maglev systems with creative writing and visual art. Students will create an advertising poster that highlights their brand new maglev operated transportation. They will design a purpose, destination and of course, the type of transportation. The goal is to convince a buyer to purchase their product!

Materials

o   Drawing paper
o   Pencils
o   Colored pencils
o   Markers
o   Scratch paper for brainstorming

Procedure

o   Students will brainstorm the function, purpose, type of transportation, and "look" of their maglev system.

TABLE 5.2, *CONTINUED*

o Student create an advertising poster that:
   • tells the viewer the type of transportation (e.g., train, car, bike);
   • tells the viewer the purpose of the transportation (what is it carrying and why?);
   • shows the viewer what it looks like;
   • shows the benefits of using their maglev over someone else's; and
   • grabs the audience's attention!

A Pull to the Past
Students will combine their knowledge of maglev systems and creative writing to write a letter to someone who lived more than 100 years ago. Students will have to decide who they are writing to and why. Why does someone in the past need to know that maglev systems exist today? How will you explain how a maglev system works? Remember, trains DID exist 100 years ago. So WHY did we invent a maglev system for a train over an engine-based train?
Materials
o Scratch paper for brainstorming o Paper for their letter *(Suggestion: Use a heavy weight so it is presentable when finished)*
Procedure
o Brainstorm the questions you should answer in your letter. You should definitely address the questions above. o Write your letter with neat and presentable handwriting.

SUGGESTIONS FOR INTEGRATING ENGINEERING AND THE ARTS IN GIFTED PROGRAMS

By working collaboratively with the arts specialist at your school, teachers can develop ways that art could enrich existing engineering lessons. Linda Gohlke of San Diego Unified School District suggested the following ways to integrate the arts and engineering (as mentioned in Saraniero, n.d.):

- Brainstorm different art forms and determine how each art form could enhance student learning. Sample ideas could consist of the following:
  - » Employ dance or movement to illustrate engineering concepts that are abstract or cannot be seen.
  - » Use improvisation or pantomime to practice oral language skills.
  - » Create visual art to help students have a conceptual understanding of content.

TABLE 5.3
Suggestions for Integrating Arts and Circuits

Overview
Students will combine their knowledge of circuits and switches with visual arts and art history in this van Gogh-inspired "The Starry Night" project.

Materials

- o   Assortment of blue, yellow, orange, black, and grey paint (tempera or acrylic)
- o   Drawing paper (computer paper is not thick enough to support this project)
- o   Hole punch
- o   Cup of water and brush for <u>each</u> student
- o   Paper plate for student paints (to serve as a palette)

Procedure

The following resource provides directions on how to create paper circuits: http://www.instructables.com/id/StarryNight-Paper-Circuits-and-Astronomy-for-Kids/#_a5y_p=3344501

Students will do the following steps once their circuits and switches are complete with the appropriate amount of LED lights for the student's constellation:

- o   Sketch constellation onto drawing paper by drawing small circles where their stars will be placed.
- o   Use a small hole punch (1/16 or 1/8 diameter, depending on your lights) to punch holes where the stars are placed.
- o   Paint your starry night by using dashed (or broken lines—the lines that you see in the middle of the road to denote that you may pass) lines.
- o   Teacher: Reference the artwork, *The Starry Night* by Vincent van Gogh from 1889. Point out the swirling movement of van Gogh's lines that create a ripple effect. The artist is trying to convey luminance. This information might help the students visualize where to place their lines.
- o   Tape your copper wire (circuits and LED lights to the back of your artwork.
- o   Push the LED lights from the back of the project, through the paper and to the front.

Storyboarding the Electrical Journey
In this project, students will combine their creative writing and visual art skills with their concept knowledge of circuits. Students will personify the electrical current as it travels through the circuit board to reach its destination. Their story will come to life as they draw their narrative by using a storyboard format.

Materials

- o   Scratch paper for literary and visual brainstorming
- o   Pencils and your desired choice of drawing medium (crayon, colored pencils, drawing pencils; suggestion: avoid using markers. They tend to "bleed" and the students are not able to achieve a high level of detail.)
- o   6" squares for storyboards
- o   Black construction paper (preferably 12"x16")
- o   Scissors and glue

TABLE 5.3, *CONTINUED*

Brainstorming Procedure
o Brainstorm (What will your personified electrical current look like? How will it act? What will it do to show you it moving through the circuits? What are they feeling? Where are they headed?) o Write a short story about your electrical current. *Suggestion: Have a set amount of details you want your students to include so that they have enough to choose from when it comes time to storyboard their narrative.*
Storyboarding Procedure
o Choose the amount of "scenes" you would like your students to produce. o Students will choose clips or scenes from their story that fit the quantity. It is important to remember that the scenes you choose MUST represent the WHOLE story. For example: You wouldn't choose the first 4 scenes because you wouldn't know the ending. o Students will draw their scenes with attention to detail. o Cut and paste the "scenes" in proper sequence onto the black construction paper. o For more information on how to use storyboarding, check out the following website: http://www.scholastic.com/teachers/article/what-are-storyboards.
Additional Idea
Here is another idea if you would prefer to incorporate dance into your lesson instead of the visual arts. In this lesson, students would use circuits to design a dance pad that buzzes or lights up when students step on it: http://teachers.egfi-k12.org/lesson-dance-pad-mania.

- Examine your engineering curriculum and determine which art forms would best enhance or provide a deeper understanding to the content. (para. 6)

Silverstein and Layne (2010) offered the following suggestions for incorporating the arts into an existing curriculum:

- use components of constructivism throughout the instruction (e.g., student-centered classroom, problem-based learning, authentic learning);
- actively engage students in constructing knowledge;
- have students create art forms that demonstrate their understanding of content;
- have students create original products that illustrate their understanding of content;
- make connections between an art form and a particular subject area; and
- ensure objectives for the arts and the particular subject area are purposeful and relevant. (pp. 2–7)

TABLE 5.4
Suggestions for Integrating Arts and Environmental Engineering

Key Questions
o How does environmental engineering relate to the arts? o Who are the artists who have been influenced by environmental engineering, oil spills, and the response to cleaning oil spills? o How can we make the engineering efforts from the oil spills relevant to our students through the lens of art?

Articles
http://www.theecologist.org/News/news_analysis/1149771/artist_attempts_take-over_of_bp_with_sales_of_oil_spill_art.html (Connections: Economics, Art, Literacy, Science) *Educational Connection:* Repurposing, up-cycling. How can we incorporate the idea of repurposing in our own artwork?

Artworks/Exhibitions

o http://inhabitat.com/presenting-mired-in-the-bayou-an-inhabitat-oil-spill-exhibit-nyc
  • Mixed media artworks (e.g., found objects, remnants from the oil spill)
  • *Educational Connection:* Exhibition located in NYC that brings the centralized destruction of the harbor across the U.S. Do you think that different areas of the U.S. viewed this differently? Regional connections? Is there something happening in our local environment? Pollution? Can our issues relate to other issues in other cities, states? What are some symbols we could use to tell this story? Students can use these symbols to create narrative art (artwork that tells a story).
 	
o http://curatorsintl.org/intensive/proposal/catalyst_artists_of_southern_louisiana_respond_to_the_gulf_oil_crisis
  • *Educational Connection:* Art created by the artists affected by the disaster. How does this change our view of the artwork? How does this change their relationship to the art? Create environmental art: art using found objects with commentary on an environmental issue.
 	
o http://jonathanferraragallery.blogspot.com/2013/09/beautiful-decay-features-generic-art.html
  • Re-imaging (Art terminology: *juxtaposition*) classic paintings by poignant and influential artists (e.g., Caravaggio, Goya, Marat)
  • *Educational Connection:* Historical: Why do you think these artists are taking historical art examples and applying them to present day issues? Can our issues today reflect the issues of our past? What environmental problems do we face today that have been happening for centuries? Students can use these symbols to create narrative art (artwork that tells a story).

KEY COMPONENTS OF ENGINEERING INSTRUCTION

CONCLUSION

Integrating the arts and engineering encourages critical thinking, provides authentic learning, supports creativity, and enhances problem-solving skills. Since the innovation and creativity that occurred during the Renaissance era, when there were no boundaries between art and science, our society has come full circle to promoting interdisciplinary learning that supports and encourages innovative thinking.

REFERENCES

Alias, M., Black, T. R., & Grey, D. E. (2002). Effect of instructions on spatial visualization ability in civil engineering students. *International Education Journal, 3,* 1–12.

The Engineer. (2006, October 12). Leonardo da Vinci. *Engineering.com.* Retrieved from http://www.engineering.com/Library/ArticlesPage/tabid/85/ArticleID/34/Leonardo-da-Vinci.aspx

Knighten, L. (2016, March 24). *Chromatography for kids activity.* Retrieved from http://www.education.com/activity/article/Color_Science_kindergarten

LaMore, R., Root-Bernstein, R., Root-Bernstein, M., Schweitzer, J. H., Lawton, J. L., Roraback, . . . Fernandez, L. (2013). Arts and crafts: Critical to economic innovation. *Economic Development Quarterly, 27,* 221–229. doi:10.1177/089 1242413486186

Left Brain Craft Brain. (2015, March 29). *Five minute craft: Magnet painting.* Retrieved from http://leftbraincraftbrain.com/2015/03/29/five-minute-craft-magnet-painting

Lynch, P. (2007). Making meaning many ways: An exploratory look at integrating the arts with classroom curriculum. *Art Education, 60*(4), 33–38.

Meethumalu. (n.d.). *StarryNight: paper circuits and astronomy for kids!* Retrieved from http://www.instructables.com/id/StarryNight-Paper-Circuits-and-Astronomy-for-Kids

Museum of Science, Boston. (2016). *Engineering is Elementary.* Retrieved from http://www.eie.org

National Academy of Engineering. (2004). *The engineer of 2020: Visions of engineering in the new century.* Washington, DC: National Academics Press.

Orhun, E., & Orhun. (2013, September). *Creativity and engineering education.* Paper presented at 41st European Society for Engineering Education Conference, Leuven, Belgium.

Robinson, C., & Baxter, S. C. (2013, June). *Turning STEM into STEAM.* Paper presented at 120th American Society for Engineering Education Annual Conference and Exposition, Atlanta, GA.

Root-Bernstein, R. S., Allan, L., Beach, L., Bhadula, R., Fast, J., Hosey, C., & Weinlander, S. (2008). Arts foster scientific success: Avocations of Nobel, National Academy, Royal Society, and Sigma Xi members. *Journal of the Psychology of Science Technology, 1*(2), 51–63.

Saraniero, P. (n.d.). *Growing from STEM to STEAM: Tips to team up the arts and sciences in your classroom.* Retrieved from https://artsedge.kennedy-center.org/educators/how-to/growing-from-stem-to-steam

Silverstein, L. B., & Layne, S. (2010). *Defining arts integration.* Retrieved from http://www.kennedy-center.org/education/partners/defining_arts_integration.pdf

Sniderman, D. (2012, April). *Leonardo da Vinci.* Retrieved from https://www.asme.org/engineering-topics/articles/history-of-mechanical-engineering/leonardo-da-vinci

Vuksanovich, B. D., & Wallace, D. R. (2011). *Evaluation of STEM + Art collaboration for multidisciplinary engineering technology laboratory.* Paper presented at American Society for Engineering Education Annual Conference and Exposition, Vancouver, BC, Canada.

SECTION 2

Integrating Innovations Into Engineering Curriculum and Instruction

 DOI: 10.4324/9781003234951-8

CHAPTER 6

Redefining Curriculum Through Engineering Practices

Using 3-D Printing for Learning

Jason Trumble

A dull cardboard box addressed to a second-grade teacher arrived in the office of an elementary school one day. Upon opening the box, the teacher's heart began to race and excitement showed on her face. She carefully removed the packaging, revealing a shiny new 3-D printer. Now she could start her Makerspace. Now her students could create. Now she could print in three dimensions. As she downloaded the software, she began to dream of the things she would teach with the 3-D printer. Her excitement and optimism soon fizzled as she tried to imagine how she could connect the 3-D printer to her second-grade curriculum.

 DOI: 10.4324/9781003234951-9

This scenario happens often. Teachers find tools that have potential to help students learn in new and exciting ways, but then struggle to truly integrate learning with the technology. The maker movement and digital fabrication in education is beginning to boom, and there are clear connections to the Next Generation Science Standards (NGSS Lead States, 2013) and engineering practices that need to be explored (Katterfeldt, Dittert, & Schelhowe, 2015). Prices for 3-D printers fall every day as schools, libraries, and communities purchase digital fabrication tools in hopes that they will help students learn, be creative, and develop as burgeoning engineers, scientists, and life-long learners. Teachers then take the mantle of incorporating these devices into curriculum, but the addition of devices can make this task difficult and unnatural. The goal of this chapter is to suggest that instead of adding technology, specifically digital fabrication, teachers look to redesign student learning as they engineer unique experiences for students to develop deep understandings through science and engineering practices.

This chapter is intended to provide practical applications for practitioners as they redefine learning for high-ability learners through innovative uses of 3-D design and printing technologies. Some rationale and explanation will be provided. Two vignettes of classroom application that incorporate 3-D design and printing will be provided.

WHAT IS 3-D PRINTING?

3-D printing has been in existence since the mid 1980s (Toutwine, Lopez, & Tweel, 2014); however, recent advances in technology have allowed for this technology to become available for the general public, and as mentioned before, the cost of both printers and supplies is decreasing as new 3-D printing manufacturers come into the market. Previously, this type of technology was only available to large businesses, which would use digital fabrication in the process of engineering and manufacturing (Toutwine et al., 2014).

Although there are multiple types of 3-D printing devices, the most popular and the one featured in the examples below is extrusion printing or fused deposition modeling. In this type of 3-D printing, a medium, usually plastic, is heated and fed through an extruder and is laid down in thin layers on a platform. The layers fuse to form a three-dimensional object.

To print a 3-D object, it must be designed, created, and rendered digitally. This is usually done with computer-aided design (CAD) software. There are many CAD programs available, both for purchase and for free online use. The specific software programs used in the classroom examples in this chapter are free and age-appropriate for students. When an object is created digitally through a CAD program, it is exported as a .stl or .obj file. This file is then uploaded to proprietary software developed by the 3-D printer manufacturer. This software digitally slices the object into thin layers, and this data is communicated to the printer, where the layers are then extruded to create the physical form of the digitally designed object.

REDEFINING TEACHING AND LEARNING

Learning solely through lecture is a thing of the past. Because rote information is at the students' fingertips, the NGSS (NGSS Lead States, 2013) propose a change in the way that students learn. Instead of individualized standards and disciplinary autonomy, the NGSS integrate concepts with an emphasis on student engagement in the inquiry process. Teachers should not teach isolated information; rather, they should plan integrated learning experiences that support student engagement and inquiry. In essence, the NGSS recommend that students become scientists and engineers as they engage in learning concepts and theories. Assessing the level at which students are engaging in the inquiry process is supported by models that propose a redefinition of learning in a digital age.

Puentedura (2006) developed the SAMR model, which is considered a viable tool for identifying levels of technology use in teaching and learning (Romrell, Kidder, & Wood, 2014). This model identifies four levels of teaching with technology—Substitution, Augmentation, Modification, and Redefinition. The lowest level, where digital technology serves as a direct substitute for an analog technology, is the substitution level. At the augmentation level, digital technologies allow for students to engage in the lesson with some functional improvement, but the task or learning experience is fundamentally unchanged. At the modification level, the task or learning experience becomes modified because of the influence of technology. Finally, at the redefinition level, the task or learning experience is completely new and would not be possible without the advent of digital technology (Puentedura, 2006). This popular model encourages the creation of new learning experiences because of technology.

As teachers assess their use of technology in the classroom through the SAMR model (Puentedura, 2006), they are encouraged to move up the levels and to veer away from traditional didactic instruction to a more student-centered inquiry model that works in alignment with the NGSS (NGSS Lead States, 2013). The substitution and augmentation levels of the SAMR model limit the curriculum to teacher-led instruction that may engage students in traditional curriculum. On the contrary, modification and redefinition levels correspond with the design of project-based learning (PBL) and the NGSS science and engineering practices. This encourages teachers to look to redefine the learning experiences they offer students.

NGSS Science and Engineering Practices

The NGSS science and engineering processes are similar to Gold Standard PBL (Larmer & Mergendoller, 2015) because they both promote problem solving and focused inquiry. The NGSS define eight practices of science and engineering that are essential for all students (National Research Council, 2012):

1. asking (for science) questions and defining problems (for engineering),
2. developing and using models,
3. planning and carrying out investigations,
4. analyzing and interpreting data,
5. using mathematics and computational thinking,
6. constructing explanations (for science) and designing solutions (for engineering),
7. engaging in argument from evidence, and
8. obtaining, evaluating, and communicating information. (p. 3)

Engineers begin their work by defining a problem that needs to be solved (NGSS Lead States, 2013). The process moves through several iterations of inquiry, where the engineer is researching and gathering information that relates to the problem or question. Models are eventually created, tested, critiqued, and shared in collaboration with others in order for improvement to occur. This can happen several times. Finally, a public product or construction is developed that answers a question or solves the problem (NGSS Lead States, 2013). Encouraging high-ability learners to engage in this process is essential. As our students become inquirers and doers, they become engineers of their own learning.

Similarly, Gold Standard PBL (Larmer & Mergendoller, 2015) consists of eight essential elements that govern the inquiry-learning experience:

1. key knowledge, understanding, and success skills;
2. challenging problem or question;
3. sustained inquiry;
4. authenticity;
5. student voice and choice;
6. reflection;
7. critique and revision; and
8. public product.

Similar to the NGSS engineering practices, the essential elements of Gold Standard PBL are not a linear model. Projects that are student centered start with an overarching problem or question in the same way an engineer has a problem that needs solving. Through inquiry, product development, communication, and collaboration, students solve the problems and produce a product that either contributes to the solution or promotes the solution to the problem. The design principles of PBL and the practices of engineering collide as teachers introduce authentic NGSS learning experiences to the classroom. New tools such as 3-D design software and printers provide new avenues for students to engage in these processes.

INCORPORATING 3-D PRINTERS IN THE CLASSROOM

There are a variety of ways to incorporate 3-D printers into the classroom for high-ability learners. Shealer and Shealer (2014) used 3-D printing in a multigrade design project. In the project, eighth graders consulted with first graders on home design, and then over the course of a few weeks, the eighth graders created and printed the homes to create communities for the first-grade students. In the process, the eighth graders considered ecology and the environment in their design and execution of the buildings (Shealer & Shealer, 2014).

In a yearlong self-inquiry, Brown (2015) observed a learning curve as a learner herself, engaging in the process of 3-D printing. Brown reported that the learning curve had three stages—beginner, designer, and advanced. A beginner would conduct a print trial, downloading and printing an existing design. At the second stage, the user conducts a design experiment, attempting to design an object that is of no particular use. At the advanced stage, the user conducts an engineering test, attempting to solve a problem by designing a solution in a CAD program and printing an object, possibly multiple times, which contributes to solving the problem. Brown (2015) printed a predesigned dragon at the beginner stage, designed a small treasure chest at the intermediate level, and finally, printed earrings at the advanced level, solving her self-perceived problem of need for accessory (Brown, 2015). As teachers incorporate 3-D design and printing into the classroom, it is crucial that they experience the technology and be prepared for challenges in the printing process. This comes through experiencing the technology prior to implementing the engineering design process as a learning activity. Additionally, the teacher must develop the ability to scaffold and support students in their struggles as they move through the learning process.

Just as Brown (2015) experienced a learning curve, high-ability students who enter the engineering and PBL process will have a similar experience. On a school visit to a high school robotics club, I found that students would use their 3-D printer to manufacture parts that improve the function of their robots. The evidence of the engineering process and the 3-D printing learning curve was evident. The students in the club wanted to manufacture a plastic claw for their robot, functionally improving its ability to grab objects. In my observation, I found four iterations of the claw, each with a flaw that prevented it from being used on the robot. The club sponsor explained to me that they planned to print an improved design, and they hoped it would work perfectly.

There are many ways for 3-D printing to be utilized in the PBL and engineering process to redefine learning for high-ability learners. As students and teachers embrace the NGSS engineering practices and PBL, the use of 3-D design and printing will present itself as an invaluable tool. Below are two vignettes of 3-D printers and 3-D design being utilized in the classroom. As examples, they are not perfect, but they provide a glimpse into how 3-D design and manufacturing can meet the NGSS engineering practices and the Gold Standard PBL design elements.

Classroom Application: Vignette 1

Students in a first-grade class learned the properties of three-dimensional shapes, including cubes, rectangular prisms, cones, pyramids, and cylinders. Students were able to identify the shapes and begin to develop vocabulary to describe the shapes. They learned to identify the faces, vertices, and edges of the shapes. To challenge the high-ability learners, the teacher decided to use the free online CAD program Tinkercad (https://www.tinkercad.com). She demonstrated how to navigate the site, drag and drop items, and manipulate items that are on the plane. The *plane* in a CAD program is the presentation of a flat surface indicating the dimensions of the X-axis and the Y-axis; additionally, the Z-axis rises vertically off of the plane. The teacher asked the students to use the geometric tools to place certain 3-D figures on the plane, manipulate them, and finally, delete them (see Figure 6.1). When the teacher was confident in the students' abilities to manipulate the program, she asked them to use the figures to create a house (see Figure 6.2).

Building the house allowed the students to engage in the design process that is essential to engineering practices. The students were able to ask questions and experiment with the design technology, as well as plan a model that connected the learning objectives with the NGSS practices for science and engineering. Using the software, the students began to recognize the elements of scale and spatial

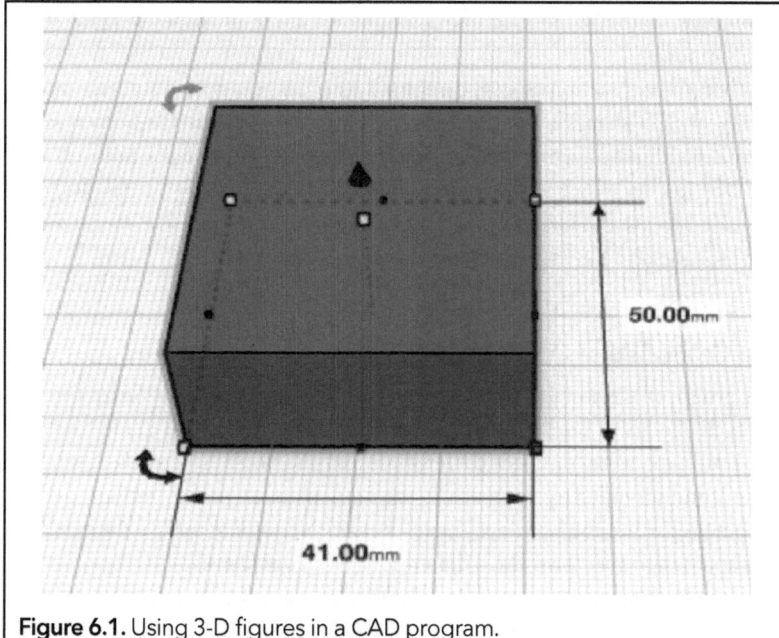

Figure 6.1. Using 3-D figures in a CAD program.

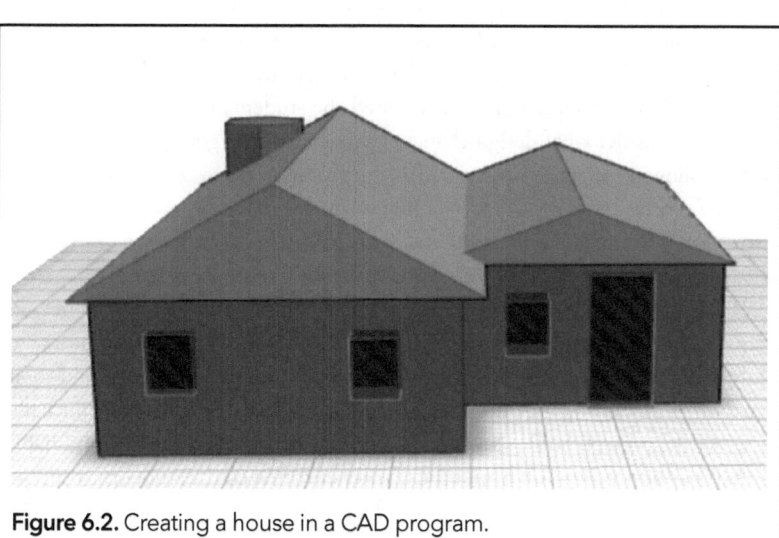

Figure 6.2. Creating a house in a CAD program.

relationships, as they connected and manipulated the 3-D shapes to build their virtual homes.

When the virtual homes were completed, the teacher assessed student understanding of 3-D shapes by engaging in a conversation about the homes using academic vocabulary, having students name shapes and count faces, vertices, and edges. The teacher then had the students download their creations as .stl files, and she decided to use the library's 3-D printer to print the houses that the students created for them to have tangible evidence of their learning.

Classroom Application: Vignette 2

A group of sixth-grade students were reading the novel *The Secret Garden* by Francis Hodgson Burnett (1911). The teachers of these sixth-grade students collaborated to create a project centered around the book. The teachers wanted to take a multidisciplinary approach to the curriculum by incorporating a design approach to parts of the learning. The teachers formed a plan to have the students create and design their own three-dimensional secret gardens.

As the students read the novel in their literacy class, they reflected on the rich language and the visualization the author provided. As they went to their math class, the teacher inquired about which aspects of the garden could be quantified and analyzed. The students and teacher discussed the garden from the text at length, analyzing the design aspects of the garden presented in the text. The math teacher pulled up pictures of gardens throughout the world and continued the discussion of garden design. This pointed the students to envision their own garden and consider what design elements make a garden "good."

The following scenario was created as the problem: *An eccentric billionaire has contracted you to design a garden. This billionaire has set a budget of $500,000 for construction, plants, sculptures, paths, and any other items in the garden. The garden can be any shape, but it must be designed to be no more than 50 meters by 50 meters.*

The students were grouped in pairs. Each pair began to think about and discuss design and the elements of a good garden design, and as they began to research and consider the text of the novel, the pictures they had seen, and their discussion, questions arose, and the students began to delve deeply into the nuance of the project. Some students questioned the location, biomes, and habitats that may affect the garden, and the teacher encouraged the students to choose an environment for their garden, which affected the choices they would make in design and selection of vegetation and other aspects.

Another group asked about irrigation and strategies for keeping the plants alive. This led to a class discussion about water features within the gardens. Subsequently, all groups decided to include a water feature. Throughout the process of the project, students were engaged in inquiry and solving problems as they arose. In this process, the students were meeting the NGSS guidelines for qualitative and quantitative inquiry as they gathered information and synthesized it into their creation.

As the project progressed, the students were provided with drafting paper in order to create a two-dimensional mock-up of their proposed garden. Each group drafted their garden to scale and presented their gardens to the other groups. Each class member was responsible for asking one question about the design of each garden, which prompted the students to consider revisions.

When the students decided they had a final version of the garden, they were allowed to use a computer to create it. The students used the CAD program Sketchup (http://www.sketchup.com) to render digital representations of their gardens. During the process of researching and creating their digital gardens, the students found items online at a retail hardware warehouse, but these items were not easily imported into the CAD software. Some students decided to exclude the items from their gardens, and some attempted to recreate the items digitally.

Continuing the engineering design process and PBL, the students and teachers collaborated and critiqued the digital gardens and proposed revisions. After repeating the critique and revision process, students rendered their gardens and attempted to print the gardens at a one-hundredth scale in order to provide a tangible representation of the garden in their presentation to the billionaire.

HOW TO PREPARE

These vignettes and the previous examples highlight a few ways that 3-D design and printing can be incorporated into curriculum. This is not an easy task. Brown (2015) discussed the learning curve and difficulty of engaging in 3-D printing and the time that it can take to learn the nuances of these complex devices. As mentioned above, it is crucial that the teacher be able to scaffold the technology skills for students. To do this, the teacher must experience the processes of 3-D design and printing. The machines are complex and sensitive, and understanding them fully before allowing students to use them is essential.

The most powerful aspect of this process is the designing of objects using CAD software. Although the printer creates a tangible object, the manipulation and iterations all happen within the software. CAD software allows for minimum risk in the process of designing. As teachers prepare to integrate these technologies into their classrooms, it is imperative that they practice and fail at creating objects, and then do it again. They must go through the process of learning that Brown (2015) defined.

This learning through doing is an invaluable experience for high-ability learners. As they traverse the process of engineering and building objects through CAD software, students gain experience with spatial manipulations, tessellations, and scaling. Students who engage with 3-D manipulations can increase their spatial understandings (Baki, Kosa, & Guven, 2011). Younger students learn about proportion and basic understanding of three-dimensional objects, as shown in the first vignette. Through the process of 3-D design and printing, students learn to solve problems, including how the printer prints objects and the effect of gravity on the design of objects. The experience of creating through CAD software can help students connect design to the physical world through 3-D printing. With these tools, students are limited only by their imaginations.

CONCLUSION

In each vignette, 3-D design through CAD software and 3-D printing were vehicles for students to engage in the engineering design process synchronous with school curriculum. The students were encouraged to consider the aesthetics and function of their creation in both situations. Although not all students will become engineers, the use of 3-D design and printing at all ages can contribute to both academic advancement and personal fulfillment (Bull, Haj-Hariri, Atkins, & Moran, 2015). Redefining how students learn through technology is an essential piece of developing high-ability learners, and 3-D design and printing is one way students can engage in this redefinition using engineering processes to solve problems.

3-D printers seem to be following a similar path as the personal computer did in the early 1980s (Eisenberg, 2013). As printers begin to pop up in different places and in different shapes and sizes, it is important that teachers develop a common understanding and vocabulary (Brown, 2015), as well as innovative ideas for integrating learning with engineering design practices.

REFERENCES

Baki, A., Kosa, T., & Guven, B. (2011). A comparative study of the effects of using dynamic geometry software and physical manipulatives on the spatial visualisation skills of pre-service mathematics teachers. *British Journal of Educational Technology, 42,* 291–310.

Brown, A. (2015). 3-D printing in instructional settings: Identifying a curricular hierarchy of activities. *TechTrends, 59*(5), 16–24.

Bull, G., Haj-Hariri, H., Atkins, R., & Moran, P. (2015). An educational framework for digital manufacturing in schools. *3-D Printing and Additive Manufacturing, 2*(2), 42–49.

Eisenberg, M. (2013). 3-D printing for children: What to build next? *International Journal of Child-Computer Interaction, 1,* 7–13.

Katterfeldt, E., Dittert, N., & Schelhowe, H. (2015). Designing digital fabrication learning environments for Bildung: Implications from ten years of physical computing workshops. *International Journal of Child-Computer Interaction, 5,* 3–10.

Larmer, J., & Mergendoller, J. (2015, May 11). *Why we changed our model of the "8 essential elements of PBL."* [Web log post]. Retrieved from http://bie.org/blog/why_we_changed_our_model_of_the_8_essential_elements_of_pbl

National Research Council. (2012). *A framework for K–12 science education: Practices, crosscutting concepts, and core ideas.* Washington, DC: The National Academies Press.

NGSS Lead States. (2013). *Next Generation Science Standards: For states, by states.* Washington, DC: The National Academies Press.

Puentedura, R. (2006, November 28). *Transformation, technology, and education in the state of Maine* [Web log post]. Retrieved from http://www.hippasus.com/rrpweblog/archives/000018.html

Romrell, D., Kidder, L. C., & Wood, E. (2014). The SAMR Model as a framework for evaluating mLearning. *Journal of Asynchronous Learning Networks, 18*(2), 1–15.

Shealer, R., & Shealer, M. (2014). Making it real: A cooperative, multigrade, 3-D design project. *Technology and Engineering Teacher, 74*(2), 8–11

Toutwine, C., Lopez, L., & Tweel, C. (2014). *Print the legend* [Motion Picture]. USA: Audax Films.

CHAPTER 7

Computer Science, Coding, and Project-Based Learning for Engineering Instruction

Irene Lee and April DeGennaro

Computer science is a promising area for the design and development of curriculum and instruction for gifted and advanced learners, particularly at the elementary school level. Computer science education is well-suited for project-based learning (PBL) and counters a fixed mindset that posits intelligence or intellect determines one's success (Dweck, 2006). The practice of creating computer programs embeds students in an engineering design process that includes developing, testing, and debugging (identifying and removing errors in computer code). Rarely, if ever, is one's first attempt at creating computer code successful. Rather, through persisting in problem solving one can achieve success.

 DOI: 10.4324/9781003234951-10

Computer science can be integrated in core subject areas (math, science, language arts, social studies) as well as connected to many areas of the curriculum. Project-based learning is a vehicle through which engineering practices and computational thinking are integrated into core subject areas. The National Research Council's (NRC; 2012) *A Framework for K–12 Science Education* and the Next Generation Science Standards (NGSS Lead States, 2013) stipulate that modern education in science should interweave content, modern scientific practice, and crosscutting concepts. In the NGSS, computational thinking is presented as one of the core scientific and engineering practices, and engineering design is well represented as both a disciplinary core idea and science and engineering practice.

Various curricula have successfully integrated computer science, engineering design, and core disciplinary content to produce learning activities suitable for all students and provide fertile ground for gifted and advanced learners to explore and grow. In this chapter, we describe three such curricula and discuss issues surrounding their implementation: LEGO's online lessons (https://education.lego.com) and Code.org and Project GUTS. LEGO's online lessons incorporate engineering design and science concepts of simple machines. The Code.org and Project GUTS *Computer Science in Science* curriculum demonstrates how computational thinking, computer science, and science education can be interwoven with an engineering design approach.

BACKGROUND

Computer science is a foundation of the practice of science and engineering, which directly contributes to the advancement of every science field. Science and engineering, in turn, push the boundaries of computer science, leading to a virtuous cycle of growth, in which advances in any of the areas reinforce the connection between all three. Computational approaches are dramatically increasing our understanding of the world and ourselves, from particle physics to biological and social systems to Earth systems science (National Science Foundation [NSF], 2014). In particular, scientists' ability to simulate natural or designed phenomena has opened up new opportunities to gain understanding and potentially solve problems within scientific disciplines, contributing new innovations, breakthroughs, and knowledge. In the 2005 President's Information Technology Advisory Committee's Report to the President, *Computational Science: Ensuring America's Competitiveness*, this computational approach was called the "third pillar of scientific practice," joining the two classical approaches of theoretical and experimental science. The NRC's *A Framework for K–12 Science Education* (2012) reflects this modern view of science. Using computational science methods, young scientists are able to augment traditional scientific practices by designing and testing computer models, and running multiple *what if* simulations on the computer that are too expensive, too dangerous, or too time consuming to run in the real world.

Engineering is defined as any engagement in a systematic practice of design to achieve solutions to particular human problems (NRC, 2011). Engineering design is a process used in the design and development of computer models and programs. The NGSS include engineering design as an essential element of science education. This addition helps teachers and students see how science and engineering are instrumental in addressing major challenges that confront society today and enable students to engage in and aspire to solve major societal and environmental challenges, such as generating sufficient energy, preventing and treating diseases, maintaining supplies of clean water and food, and global environmental change.

The engineering design process is a series of steps that engineers follow to solve a problem. Many times, the solution involves designing a product that meets certain criteria and/or accomplishes a certain task (Museum of Science, Boston, 2016). Engineers do not always follow the steps in order. It is very common to design something, test it, find a problem, and then go back to an earlier step to make modifications to the design. This is called *iteration refinement*. In the NGSS, students are expected to be able to define problems—situations that people wish to change—by specifying criteria and constraints for acceptable solutions, generating and evaluating multiple solutions, building and testing prototypes, and optimizing a solution. In developing computer programs (coding), students engage in the engineering design process; they design, test, and refine algorithms as solutions to problems. The following sections provide strategies for supporting gifted and advanced learners during implementation of computer science education and example curricula for incorporating the engineering design process.

COMPUTER SCIENCE EDUCATION FOR GIFTED AND ADVANCED LEARNERS

PBL in computer science affords many opportunities to extend learning and avoid boredom. When students participate in computer science activities, where they design, build, and test creations, they are challenged to take on the roles of practicing engineers engaged in an iterative refinement cycle (Tomlinson et. al, 2002). Debugging code, analyzing designs for performance errors, and collecting feedback data from testing their design are authentic engineering practices in which gifted students can engage. The rigor of the practice and relevance of addressing real-world problems provide motivation for engagement.

Striking the right balance of on-computer time with hands-on design and build activities is central to supporting the development of engineering practices. Young engineers are capable of using technology to design and develop creations. Classrooms that provide varying amounts and types of computer science activities (both on- and off-computer) give students opportunities to develop thinking skills, such as decomposing problems toward finding patterns and building theories that can be tested and applied. Teachers who provide students with complex problems to solve, combined with opportunities to experiment with solutions, give students practice at persisting to success. Using a facilitation model, basic skills are introduced by an instructor after which a student works creatively to identify problems, pose solutions, and persist in a process of iteration toward success. Project- and problem-based instruction lend nicely to computer science. Students can be given a problem or scenario to solve, and, with access to building materials and computer software, they can develop models and prototypes, developing the habits and discipline of practitioners in the field.

Gifted and advanced learners may experience challenges related to the interpersonal, affective, and behavioral aspects of computer science education. Carol Dweck, author of *Mindset: The New Psychology of Success* (2006), wrote that giftedness depends as much on our experiences and attitudes about goals, effort, and setbacks over time as it does innate intelligence: "Genius and great, creative contributions are the product of passion, learning, and persistence" (Dweck, 2008, para. 1). Students accustomed to getting the right answer easily and quickly and seeking perfection may get frustrated easily with the iterative development process, the speed of progress, and/or with errors encountered in developing computer programs and models. Common pedagogical practices used in computer science classrooms, such as pair-programming and peer instruction, may exacerbate difficulties gifted students have with communication, self-control, and impatience. Difficulties with attention and organization, compounded by unrealistic expectations of immediate success, can be problematic for gifted students and lead to burnout. Nevertheless, the benefits of computer science education outweigh the challenges for gifted students, and strategies exist to support student learning.

Strategies to Support Students in Productive Struggle in Computer Science

Productive struggle has been described in mathematics education literature as an effortful practice that builds useful understanding and skill by making sense of something or figuring out something that is not immediately apparent (Hiebert

& Grouws, 2007). Productive struggle is a positive student experience and often involves students developing tenacity and using good strategies. In the context of computer science, this description is relevant, but, in addition to making sense of concepts and structures and uncovering mathematical relationships, computer science learners are often creating algorithms in which their understandings of concepts and mechanisms are embedded. This actualization of their thoughts and understanding into a piece of working computer code adds another layer of practice, in which productive struggle may occur. Strategies exist that may help teachers foster productive struggle in gifted and advanced students (McCoy & Cudjoe, 2015):

- **Anticipate student struggles and misconceptions that might occur.** Plan ways to support students without removing the opportunities for students to develop deeper understanding. Get familiar with common issues/misconceptions students may have. Prepare questions and prompts that may lead students to reflect.
- **Offer process praise.** Instead of praising students for their talent or intelligence, praise students for the strategies they use, the work they do, and their persistence or effort (Kamins & Dweck, 1999).
- **Use open-ended questions.** Frame questions in a way that multiple approaches and answers are expected and accepted. Also provide multiple entry points and ways students can relate to the problem (e.g., when creating algorithms, students can solve the same problem in different ways; Prince & Felder, 2006).
- **Utilize parallel tasks (easy, medium, hard challenges).** Assign two or more tasks that are designed to meet the needs of students working at different levels, but address the same fundamental concept and use a similar context, so that they can be discussed simultaneously (e.g., challenging students to use simple looping vs. nested looping to draw a figure; Gregory, 2003; Tomlinson 1999; Tomlinson, 2001; Tomlinson & McTighe, 2006).
- **Attend to student-teacher interactions.** Consider the balance between supporting students enough to avoid frustration while maximizing learning and independence (Malone, 1981).
- **Incorporate ongoing formative assessment.** Ask students to express what they understand about new materials, and allow them to pinpoint and correct their knowledge gaps and misconceptions (Goldenberg et. al., 2015; Popham, 2008).
- **Use spaced/distributed practice** (Baddeley, Anderson, & Michael, 2009; Benjamin & Tullis, 2010; Rohrer & Taylor, 2006; Sobel, Cepeda, & Kapler, 2011). Help students to understand that they don't have to learn everything during their first attempt. Support students by circling

back to topics to deepen and refine understanding over time (Cimpian, Arce, Markman, & Dweck, 2007; Rawson, Dunlosky, & Sciartelli, 2013). This spiraling has been found to produce lasting learning because long-term memory of material is strengthened each time information is actively retrieved (Sweller, 1988).

- **Incorporate mixed practice.** Practicing different types of questions and problem solving builds learning-for-transfer more effectively (Rohrer, 2009).

EXAMPLES OF CURRICULA THAT HAVE BEEN INTEGRATED INTO K-8 STEM CLASSES TO DEVELOP ADVANCED TALENTS IN ENGINEERING

Computer Science

Code.org introduces computer science through learning to code at the elementary level with a suite of five courses, collectively called "Computer Science Fundamentals." Hadi and Ali Partovi launched Code.org in January 2013 as a nonprofit focused on making computer programming more accessible. Students in elementary school can explore the five courses of 20 lessons each. The courses blend online, self-guided, and self-paced tutorials with unplugged activities that require no computer at all. Each course could be implemented as a standalone unit or used for activities spanning a semester or school year. Even kindergarten-aged prereaders can participate.

Code.org also provides 20 hours of free online professional learning for teachers and, at the time of publication, offers free one-day workshops for K–5 teachers. These workshops are led by experienced Code.org Facilitators across the United States. The Hour of Code is a one-hour introduction to computer science, designed to demystify code and show that anybody can learn the basics of computer science. At middle school, Code.org offerings include modules to integrate computer science and algebra and computer science and science (described on p. 104). A new standalone computer science course is to be released in 2017. Two yearlong high school courses designed to broaden computer science participation are also available.

Robotics

LEGO offers apps that include lessons that link science concepts to a classic LEGO construction manual. WeDo 2.0 (https://education.lego.com/en-us/elementary/shop/wedo-2) allows children to build and program LEGO models that incorporate everyday concepts of simple machines and are geared toward K–5 settings. Children program their robots to move and make sound using software designed specifically for WeDo 2.0, and rely on a USB cable for power and transfer of programs. The logic involved in building a program helps students develop thinking skills necessary in computer science, and the focus on simple machines helps develop their understanding of engineering design principles. The WeDo software has a very simple drag-and-drop coding interface that lets students program basic functions. Or, students may ignore the instructions entirely and build and program their own robots.

LEGO MINDSTORMS (http://mindstorms.lego.com), introduced in 1998, are robots geared toward slightly older students. The EV3 version offers more power, more ports, and the addition of an SD card slot. The EV3 robot is used in FIRST LEGO League competitions discussed in Chapter 1 of this book. EV3 also lends itself to noncompetitive use in classrooms, due to the many online tutorials, LEGO-provided materials such as the Green City add-on kit (https://education.lego.com/en-au/lego-education-product-database/mindstorms/9594-green-city-challenge-set), and an easy-to-use book by Damien Key, *Classroom Ideas for the Busy Teacher: EV3*. Both WeDos and EV3s are excellent options for incorporating computer science and engineering design because students engineer and build as they code programs and run trials in a process of iterative engineering design.

Computer Modeling and Simulation

Computer modeling and simulation are powerful tools in a scientist's toolbox. Scientists develop computer models as abstractions of the real world that incorporate their theories of how the real world works, then use the models as experimental test beds and simulate how the modeled world or system might change over time. Computational thinking is the human ability to formulate problems so that their solutions can be represented as computational steps or algorithms to be carried out by a computer. In computer modeling and simulation, computational thinking entails deciding which elements and interactions to include in a model and which to leave out (abstraction), automating behaviors and encoding interactions (automation of algorithms), and determining if the abstractions made were valid (analysis; Cuny, Snyder, & Wing, 2010).

Importantly, modeling and simulation have been shown to have broad appeal and to provide students from diverse backgrounds and with a range of educational needs with opportunities to successfully engage in modern scientific practice. Project GUTS (http://www.projectguts.org) serves students from underrepresented groups in STEM, including Native Americans, African Americans, Hispanics/Latinos, and young women. In these programs, students have been able to incorporate their own reality into scientific investigations through computer modeling and simulation, resulting in deep engagement in NGSS content and practices.

In *Computer Science in Science* curriculum modules (https://code.org/curriculum/science), developed by Project GUTS (2014) in partnership with Code.org, middle and high school students are introduced to modeling and simulation through a progression of design and build activities that lead up to developing their first computer model. The progression starts with students learning how to create models by writing computer codes to:

- create agents (an object or individual in the computer model),
- direct the movement of these agents,
- instruct agents to leave trails in their environment,
- enable agents to detect and react to trails left by other agents, and
- instruct agents to interact with other agents—for example, sometimes a collision will result in a change in an agent's state.

With these five simple behavior patterns, students will have the tools they need to model and study numerous real-world phenomena, such as the spread of disease.

Agent-based modeling environments, specifically StarLogo TNG and StarLogo Nova have been used with sixth- to eighth-grade students in science classes to delve into science concepts through modeling and simulation. A progression called Use–Modify–Create provides a scaffolded approach to engage students in computational thinking within modeling and simulation environments. After an initial description of the model is presented to the student, the progression can guide the student to deeper levels of understanding of the model and the phenomena being modeled.

For example, in a life science unit on ecosystems, students are given a simple ecosystem model—a pond. The students, upon inspecting the underlying code, see that the world is made up of fish and plankton that move through the pond with a little bit of randomness in their motion. Fish lose energy when they move. When a fish encounters a plankton, the plankton gets eaten and the fish gains energy. When the fish's energy has reached a threshold, it gives birth to a new fish and gives the new fish some of its energy.

In the Use stage, students run experiments on a pre-built computer model. In this example, students are asked to find an initial number of fish that leads to a pond ecosystem where both fish and plankton can persist over time. By changing initial number of fish through a user interface widget and running repeated experiments, students might discover that there are three possible outcomes: (1) Fish die out, but the plankton survive; (2) all the plankton get eaten, and then the fish die out; or (3) both the fish and plankton populations persist in dynamic equilibrium.

In the Modify stage, students may initially want to see if cosmetic change, such as changing the color of the plankton, impacts the system's behavior. Over time, students ask deeper questions and modify the model with increasing levels of sophistication. Some examples of questions that students might pose and answer for themselves are:

- How many different patterns of population growth and death can be found? What are they?
- What is the impact of lowering the maximum number of plankton in the ecosystem?
- What happens to the fish population if it takes more energy for the fish to reproduce?
- How can I achieve a long lasting ecosystem?
- What is the impact of adding a top predator?

Answering the questions requires that students use the scientific practices promoted by the NGSS. Students engaged in *computational thinking* when they used abstraction to specify a new agent, a resource limitation, or an algorithm for an agent's behavior. They *conducted a scientific inquiry* using the model as a test bed. They *uncovered new patterns* generated by the modeled system. They *analyzed and interpreted data*, and participated in argumentation *from evidence* as they sought to validate their model by comparing it to what they know of the real world. Through a series of questions, modifications, and iterative refinements, new skills and experiences in engineering design are gained, and new scientific understandings are developed as what was once someone else's model becomes one's own.

This computational science investigation can be used to satisfy the following NGSS performance expectations: Ecosystems: Interactions, Energy, and Dynamics. The module can also be used to incorporate the NGSS engineering practices and disciplinary core ideas in the area of engineering such as: defining and delimiting an engineering problem; developing possible solutions; optimizing the design solution; interdependence of science, engineering, and technology; and influence of engineering, technology and science on society and the natural world.

SUMMARY

Early examples of curriculum and instruction at the elementary school level show promise for the integration of computer science and engineering design in STEM education. For gifted and advanced learners, PBL in computer science offers many benefits and fosters perseverance and persistence in problem solving. As computer science education gains a foothold in K–12 education, it will be important to consider ongoing systemic concerns, such as the availability of time for PBL in traditional classes and opportunities for professional learning that enable teachers to offer instruction infused with computer science and engineering design principles. As educators, we need to develop a growth mindset about assessment as a tool that provides data to help students progress on a continuum from novice to expert across K–8 settings. Assessment that mirrors the process of iterative refinement (and does not expect mastery at the first attempt) enables students to figure out what they do well and what they can do better at and is an essential part of making lasting changes to engineering and computer science education.

REFERENCES

Baddeley, A., Eysenck, M. W., & Anderson, M. (2009). Working memory. In A. Baddeley, M. W. Eysencky, & M. Anderson (Eds.), *Memory* (pp. 86–87). New York, NY: Psychology Press.

Benjamin, A. S., & Tullis, J. (2010). What makes distributed practice effective? *Cognitive Psychology, 61,* 228–247.

Cimpian, A., Arce, H., Markman, E. M., & Dweck, C. S. (2007). Subtle linguistic cues impact children's motivation. *Psychological Science, 18,* 314–316.

Cuny, J., Snyder, L., & Wing, J. M. (2010). *Demystifying computational thinking for non-computer scientists.* Manuscript in preparation.

Dweck, C. (2008, April). Giftedness: A motivational perspective. *Digest of Gifted Research.* Retrieved from https://tip.duke.edu/node/888

Dweck, C. S. (2006). *Mindset: The new psychology of success.* New York, NY: Random House.

Goldenberg, P. E., Mark, J., Kang, J., Fries, M., Carter, C. J., & Cordner, T. (2015). *Making sense of algebra: Developing students' habits of mind.* Portsmouth, NH: Heinemann.

Gregory, G. H. (2003). *Differentiated instructional strategies in practice.* Thousand Oaks, CA: Corwin Press.

Hiebert, J., & Grouws, D. A. (2007). The effects of classroom mathematics teaching on students' learning. In F. K. Lester (Ed.), *Second handbook of research on mathematics teaching and learning* (pp. 371–404). Charlotte, NC: Information Age.

Kamins, M., & Dweck, C. (1999). Person versus process praise and criticism: Implications for contingent self-worth and coping. *Developmental Psychology, 35,* 835–847.

Malone, T. (1981). What makes computer games fun? *ACM SIGSOC Bulletin, 13,* 143.

McCoy, A., & Cudjoe, K. (2015, April). *Productive struggle.* Presentation at the Annual Conference for the National Council of Teachers of Mathematics, Boston, MA.

Museum of Science, Boston (2016). The EiE curriculum. *Engineering is Elementary.* Retrieved from http://www.eie.org/eie-curriculum

National Research Council. (2012). *A framework for K–12 science education: Practices, crosscutting concepts, and core ideas.* Washington, DC: The National Academies Press.

National Science Foundation. (2014). *STEM + Computing Partnerships (STEM+C): Program Solicitation.* Retrieved from http://www.nsf.gov/pubs/2015/nsf15537/nsf15537.htm

NGSS Lead States. (2013). *Next generation science standards: For states, by states.* Washington, DC: The National Academies Press.

Popham, W. J. (2008). *Transformative assessment.* Alexandria, VA: Association for Supervision and Curriculum Development.

President's Information Technology Advisory Committee. (PITAC, 2005). *Computational science: Insuring America's competitiveness: Report to the President.* Washington, DC: National Coordination Office for Information Technology Research and Development. Retrieved from https://www.nitrd.gov/pitac/reports/20050609_computational/computational.pdf

Prince, M. J., & Felder, R. M. (2006). Inductive teaching and learning methods: Definitions, comparisons, and research bases. *Journal of Engineering Education, 95,* 123–138.

Rawson, K., Dunlosky, J., & Sciartelli, S. (2013). The power of successive relearning: Improving performance on course exams and long-term retention. *Educational Psychology Review, 25,* 523–548.

Rohrer, D. (2009). The effects of spacing and mixing practice problems. *Journal for Research in Mathematics Education, 40,* 4–17.

Rohrer, D., & Taylor, K. (2006). The effects of overlearning and distributed practice on the retention of mathematics knowledge. *Applied Cognitive Psychology, 20,* 1209–1224.

Sobel, H. S., Cepeda, N. J., & Kapler, I. V. (2011). Spacing effects in real-world classroom vocabulary learning. *Applied Cognitive Psychology, 25,* 763–767.

Sweller, J. (1988). Cognitive load during problem solving: Effects on learning. *Cognitive Science, 12,* 257–285.

Tomlinson, C. A. (1999). *The differentiated classroom: Responding to the needs of all learners.* Alexandria, VA: Association for Supervision and Curriculum Development.

Tomlinson, C. A. (2001). *How to differentiate instruction in mixed-ability classrooms* (2nd ed.). Alexandria, VA: Association for Supervision and Curriculum Development.

Tomlinson, C., Kaplan, S., Renzulli, J., Purcell, J., Leppien, J., & Burns, D. (2002). *The Parallel Curriculum Model: A design to develop high potential and challenge high-ability learners.* Thousand Oaks, CA: Corwin Press.

Tomlinson, C. A., & McTighe, J. (2006). *Integrating differentiated instruction and understanding by design.* Alexandria, VA: Association for Supervision and Curriculum Development.

CHAPTER 8

From Consumer to Producer

Gifted Education and the Maker Movement

Krissy Venosdale and Brian Housand

Imagine a classroom scattered with tables, a variety of comfortable chairs, and the hum of conversation. Small groups of students work collaboratively in exploring, inventing, questioning, and designing. Scraps of cardboard, hot glue guns, and simple power tools hang on the walls, making the space look more like a shop class or tool garage that one might remember from his or her own school experience.

One group of students works with a Makey Makey (http://www.makeymakey.com) hooked to a laptop, as they design a giant piano with aluminum foil keys that can be played with their feet. The Makey Makey makes invention possible, as students can use virtually anything that con-

 DOI: 10.4324/9781003234951-11

ducts electricity to control the computer—a banana, a bowl of Jello, a spoon, and even themselves!

Across the room, two students are excitedly working on carpet squares, closely gathered around an iPad, using the iOS app GarageBand to develop a podcast about the school's upcoming musical. Another group of students—some students sitting, some leaning over the table—is designing a holder for the classroom headphones, using a website with 3-D design software, that can then be printed on a 3-D printer.

At one table, a girl works alone on a program using Scratch (http://scratch.mit.edu). As she finishes another bit of code, she asks the students working with the Makey Makey, "Let's see if we can combine my Scratch project with your Makey Makey." These tools, powerful when used alone, gain an entirely new level of learning power when they become the pieces of a collaborative project bringing students' ideas and creativity together.

This unique classroom and the students in it, passionately working on a variety of creative projects, may sound similar to a gifted education classroom, where differentiation, social experiences, and imagination reign. This classroom, however, is filled with technology tools and the means for students to bring their own ideas to life with their own hands. It is also known as a Makerspace. The teacher rotates throughout the room, sitting briefly with groups of students and pushing thinking with big questions, such as, "What would happen if you tried this?" or "How do you know this will work?" The teacher is a facilitator to encourage, support, and push students' thinking to deeper levels.

THE MAKER MOVEMENT

Beyond what we commonly think of as technology integration, the Maker Movement has students apply the use of tools, rather than simply learning how they work. Using the tool becomes an authentic necessity when driven to solve a real problem and create a tangible product or prototype. It is the kind of learning that kids who have insatiable curiosity, quest for deep knowledge and understanding, and see connections that others cannot, thrive upon. It moves thinking away from consumption of information to the creation of new thinking, learning, and doing. Making focuses on student thinking, allowing their thoughts to lead, while teachers guide and support. Note that this is not a new idea in gifted education. In 1977, Renzulli wrote about the importance of transitioning students from consumers of facts to producers of new knowledge. In many ways, the Maker

Movement is the fruition of the original intent of the Enrichment Triad Model within a 21st-century learning environment.

One of the primary goals of the Schoolwide Enrichment Model is to "infuse into the general education program a broad range of activities for high-end learning that will challenge all students to perform at advanced levels, and allow teachers to determine which students should be given extended opportunities, resources, and encouragement in particular areas where superior interest and performance are demonstrated" (Renzulli & Reis, 2014, p. 46). It is precisely this type of teaching and learning that is the hallmark of the Maker Movement. In Makerspaces, students are provided with opportunities to investigate projects of interest, and they are provided with the necessary resources, tools, and materials to accomplish their tasks. As a part of this, students will often need a set of Type II experiences (Renzulli, 1977), or how-to training sessions. Finally, teachers and students alike provide the encouragement to build, create, make, and do something bigger and better than ever before. This encouragement to build, as well as "learning by constructing knowledge through the act of making something shareable" (Martinez & Stager, 2013, p. 21), is at the core of the Maker Movement.

Schools are beginning to shift their overall focus from consumption to creation. As this trend grows, the transition to learning experiences from simplistic use of a technological device is taking place. Learning moves toward the more advanced creation of solutions to solve authentic problems, inventions to better the world, and even the simplification of old tech deconstructed to become something more than it once was. This trend, the Maker Movement, is building on the move toward integration of STEAM (science, technology, engineering, art, and mathematics) in education and opening new doors to creativity and providing opportunity in inexpensive and readily available ways. In schools where gifted programs were once resigned to rooms at the end of the hall, were held on campuses far from where students attend their regular classes, or were severely underfunded, and where different types of learning in the classroom were far from understood, the Maker Movement breathes new life into providing challenging learning experiences and invites student voices to be heard in the development of curriculum, ideas, and the future of learning.

As the Maker Movement gains steam in education, it also gains steam around the world. Maker Faires (http://makerfaire.com) are gatherings of "makers," a collection of people including crafters, educators, tinkerers, hobbyists, engineers, science clubs, authors, artists, students, and commercial exhibitors. These festivals showcase inventions and creativity and celebrate the work of makers from the community and around the world. A colorful butterfly bicycle rides by with 6-foot wings flapping in the breeze, while a group of teenagers share the solar-powered car they built for their college engineering program. Children create with LEGOs strewn in a giant pit of creativity, while a man shares his artwork created with the fired ends of light bulbs. The unique ideas, the examples of human innovation, and the value of creativity are the exact things valued at the Maker Faire, and the things we seek to value in our gifted programs. Things like cognitive skills, technical skills, social-emotional development, and empowerment are all well served by Makerspaces (Vossoughi, Escudé, Kong, & Hooper, 2013).

HOW TO GET STARTED

A teacher wondering how to infuse the principles of the Maker Movement into his current classroom, homeschool, or any other learner setting does not have to look far to find resources. Open-source, or materials openly shared to be remixed, reused, and revamped, are one of the key components of the Maker Movement. Starting a Makerspace in your school can be a daunting task, thus the *Makerspace Playbook: School Edition* (http://makered.org/wp-content/uploads/2014/09/Makerspace-Playbook-Feb-2013.pdf) was created. However, note that a Makerspace is not just one thing, and your Makerspace may look very different than other Makerspaces. The focus should be on your students and what they are interested in making.

Through online resources, there is no shortage of tutorials, how-to websites, and DIY articles where teachers, students, and curious thinkers can get started. As the amount of resources can become overwhelming, one must remember that making is not about the stuff you buy to stock a Makerspace, but rather it is about the mindset of creating, inventing, and problem solving in a learner-led environment. With the teacher as facilitator, the student thinking drives the process and the outcome. Collaboration becomes a more natural fit, as it is authentically involved in design challenges and creation. It is this mindset that is at the core of a Makerspace, and it is worth fostering, cultivating, and providing the space for it to flourish.

THE TOOLS

The benefits of "making" run far and wide. As Halverson and Sheridan (2014) noted, "the great promise of the maker movement in education is to democratize access to the discourses of power that accompany becoming a producer of artifacts, especially when those artifacts use twenty-first-century technologies" (p.

500). Kafai, Fields, and Searle (2014) demonstrated that making can challenge our understanding of what counts as a legitimate learning activity. A "disruptive" making activity that can bring both "hard" and "soft" skills to the maker ecology, opening up our understanding of what counts as making (Halverson & Sheridan, 2014).

Many making activities can be used to get started. Coding, robotics, 3-D fabrication, circuitry, and the development of wearable technology can be utilized with learners of all ages. The variety provides endless opportunities for students to not only learn but also learn how to learn. As Mark Hatch (2014) noted, "We must learn to learn. We must develop our skills at creating, developing, and nurturing things and services that others value. The age of being a cog in a big machine and marching one's way to a defined benefit plan retirement is over" (p. 146).

Coding

Coding and programming may involve simply using online programs, such as Scratch (https://scratch.mit.edu), Tynker (https://www.tynker.com), or Hopscotch (https://www.gethopscotch.com). Younger students can get started with apps like the MIT-developed Scratch Jr. (http://www.scratchjr.org) or tools like Daisy the Dinosaur (http://www.daisythedinosaur.com). These apps teach children the basics of programming, allowing them to try sequences and develop understanding of basic coding. Students can even use TurtleArt (http://turtleart.org) to develop code in a drag-and-drop format to create a work of art. Students can manipulate the variables and explore how their art changes. The incorporation of art into coding provides another benefit to students with an insatiable desire for creative learning experiences. More advanced coders can use Arduino (https://www.arduino.cc) to develop interactive projects that are only limited by their imaginations. With the incorporation of tiny motors; sensors, including pressure, temperature, and sound; or even LED lights, students can truly engineer working solutions to change the world—without leaving the classroom.

Robotics

Although coding and programming offer a great deal to making in the classroom, robotics is another natural fit. Whether using LEGO robotics, Wonder Workshop's Dash and Dot (https://www.makewonder.com), or Sphero (http://www.sphero.com), students can develop robots that can be used with standalone iOS or Android apps to drive, move, and manipulate them through a course. They can also be used with coding apps like Tickle for iOS (https://tickleapp.

com) to allow students to program robots or even fly drones and develop coding skills in a real-world application of knowledge.

3-D Printing

Many schools have long-held robotics clubs, but an unfamiliar territory may be 3-D fabrication. As the 3-D printer becomes more affordable for classrooms and the software becomes more user-friendly, the incorporation of additive fabrication, or 3-D printing, opens students up to a world of bringing ideas into tangible forms. Students can create prototypes to solve problems or even design complex works of art using 3-D printing software. Programs like Tinkercad (https://www. tinkercad.com) provide free online software for students to design their work. Apps such as Morphi (http://www.morphiapp.com) and Blokify (http://www. blokify.com) allow students to use their touch to design and print 3-D creations. Thingiverse (http://www.thingiverse.com) is a virtual library of 3-D designs from all over the world, openly shared to inspire, print, and build on in your own classroom. The 3-D printer is being used to print everything from wrenches on the International Space Station to lost buttons for home washing machines. In the hands of a gifted learner, the possibilities of creation will know no limits.

Circuitry

There are many kits available to introduce students to circuitry. Tools like littleBits (http://littlebits.cc) empower the learner to invent with a snap-together magnetic kit that powers up right out of the box. Students can invent things that spin, light up, and make noise, and that is just within the first 5 minutes of opening the kit. The simplistic tools allow the learner to lead the discovery and use the tool to develop ideas and prototypes without soldering experience.

Even more simplistic, tiny LED lights, small nickel cell lithium batteries, and copper tape can be used by students to begin experimenting with designs they have built. The materials, inexpensive and readily available in hobby stores, allow students to develop their understanding of circuits and then begin to incorporate their own designs in more advanced ways. The youngest learners enjoy taping them to cardboard to develop light-up sculptures, houses, hats, or whatever they can dream up.

With Squishy Circuits (http://www.stthomas.edu/SquishyCircuits), salt-dough becomes a conductor to power LED lights with a battery pack. The activity invites the youngest of learners to play, experiment, and explore how electricity works, cause and effect, and creative problem solving. Students can form their own dough creations, develop paths with the conductive dough, and push LED lights into the material to light up their designs.

The Circuit Scribe (http://www.electroninks.com/circuit-scribe-conductive-ink-pen), a conductive ink pen, allows makers to draw lines on paper, cardboard, or fabric that will conduct currents from a power source. Paint, thread, and even glue are also available in conductive forms, bringing a whole new side of innovation to circuitry.

Most recently, wearable technology is an area that offers the integration of textiles, engineering, programming, and creativity into our everyday lives. With the development and popularity of smart watches and activity trackers, students are familiar with the idea of wearable technology. Other possibilities include the LilyPad wearable Arduino (http://lilypadarduino.org) to create wearable art, clothing infused with technology, or even works of art that become interactive in a whole new way. By incorporating lights and sensors in a painting or drawing, students can create flowers that spin, suns that light up, or even paintings that play music.

DESIGN THINKING

No matter the tool or avenue chosen by a maker, the process must remain the focus of the experience. Design thinking, an iterative process, allows students to connect with a problem by thinking about the needs of those involved, by brainstorming in a completely open format, and by working to use creativity, optimism, and resilience to push thinking forward in developing a solution. Making offers a natural fit for design thinking because of its roots in collaboration, problem solving, invention, and iterative processes. The pressures of finding the one right answer are gone when there are many right answers. Students think about the needs of someone with a problem, define the problem, brainstorm, create a prototype, and test their solutions. As in the real world, the first try is rarely the most powerful idea. Failure becomes a practice that is overcome again and again in working toward a solution. Whether writing a line of code, developing a Squishy Circuit sculpture, or creating a cardboard chair, the opportunities for failure present themselves, and the motivation of students to solve the issue is authentic and driven because of their personal investment in the design of the project. This practice helps prepare students not just for formal education, but also for life.

In *Creativity, Inc.: Overcoming the Unseen Forces That Stand in the Way of True Inspiration* (Catmull & Wallace, 2014), Ed Catmull, president of Pixar Animation Studios and Walt Disney Animation Studios, described the importance of failure in creating their films: "Failure isn't a necessary evil. In fact, it isn't evil at all. It is a necessary consequence of doing something new" ("Starting Points: Thoughts for Managing a Creative Career," para. 18). Many of our gifted students feel that they have to be perfect in all that they do especially when it comes to school. The Maker Movement can help free them of the trap of perfectionism and help them realize that making mistakes is a part of the process.

EVOLUTION OF A MAKERSPACE

A Makerspace may begin as simply as a table positioned at the side of the classroom, as a small spot in a library with some items to tinker with, or as an open space in the commons of a high school. There are no barriers to socioeconomic status, space requirements, or a list of things that must be obtained before starting. By providing students with real-world challenges and the tools to engineer solutions, making can take root in the learning environment by planting the seeds for design thinking and creativity.

Purchasing an endless list of expensive items to get started is not necessary. Instead, start with students' interests and ideas. Again, in gifted education, the idea of identifying students' interests can be traced back to Renzulli's Enrichment Triad Model (1977). Invite students to create, lend their voices to the design of the spaces, and develop a place where they are empowered to think, grow, create, and make.

Begin with basics such as cardboard, scissors, staplers, paper, markers, and an open space to work, talk, and think. The key in making is not the materials available, but rather the freedom of the learners to explore, invent, create, and, most importantly, do. In fostering a mindset of making, students will begin to think differently across the curriculum. Open-ended inquiry and curiosity will drive their thinking, iteration will become the new normal, and perseverance in problem solving will be fostered day in and day out. Making honors students for who they are and what they bring to the table. In doing so, we show learners that we believe in their abilities, and we contribute to an environment that supports learners' beliefs in themselves.

CONNECTING WITH THE WORLD

The beauty of making is that it is meant to be shared. With the power of technology, students can create blog posts, online guides, digital books, and videos about their creations. Collaboration with students in the same classroom, the next county, or even another state or country becomes a way to enhance the design thinking process and the student's life. The idea that you are working on a solution alone at your desk is one that many are comfortable with. But making that idea better by working with others is something that our globally connected society will not only offer, but also require for success. Pushing your thinking by asking for help across the room can enhance thinking. Working on an idea with students in a videoconference from various parts of the world? It's the type of global thinking that will prepare our students for a great future.

It all begins with a spark of curiosity, the freedom to wonder, and the time to explore. Today's students will face a future filled with challenges, problems, and issues that we can only imagine. To discover solutions, they must be able to problem solve and create. To better prepare them not for our world, but for their world, the world of the future, we must give them the opportunity to transform from consumers to producers. This is the rise of the Maker Movement.

REFERENCES

Catmull, E., & Wallace, A. (2014). *Creativity, Inc.: Overcoming the unseen forces that stand in the way of true inspiration* [E-reader version]. New York, NY: Random House.

Halverson, E., & Sheridan, K. (2014) The maker movement in education. *Harvard Educational Review, 84,* 495–504.

Hatch, M. (2014). *The maker movement manifesto.* New York, NY: McGraw-Hill.

Martinez, S. L., & Stager, G. S. (2013). *Invent to learn: Making, tinkering, and engineering in the classroom.* Torrance, CA: Constructing Modern Knowledge Press.

Renzulli, J. S. (1977). *The enrichment triad model: A guide for developing defensible programs for the gifted and talented.* Mansfield Center, CT: Creative Learning Press.

Renzulli, J. S., & Reis, S. M. (2014). *The schoolwide enrichment model: A comprehensive plan for educational excellence* (3rd ed.). Waco, TX: Prufrock Press.

Vossoughi, S., Escudé, M., Kong, F., & Hooper, P. (2013). *Tinkering, learning & equity in the after-school setting.* Paper presented at the FabLearn III Digital Fabrication in Education Conference, Stanford, CA. Retrieved from http://fablearn.stanford.edu/2013/wp-content/uploads/Tinkering-Learning-Equity-in-the-After-school-Setting.pdf

SECTION 3

Designing Engineering Curriculum for High-Ability Learners and Assessing Student Performance

 DOI: 10.4324/9781003234951-12

CHAPTER 9

Integrating Engineering Design Processes Into Classroom Curriculum

Michelle B. Buchanan and Debbie Dailey

The emphasis placed on engineering by the Next Generation Science Standards (NGSS Lead States, 2013) necessitates that general and gifted education teachers adapt their curriculum to integrate engineering design practices. The NGSS recommend the following progression of engineering practices for students: Students in the early grades are presented with problems and test and compare possible solutions to these problems. As they progress in grade levels, students identify the problem and develop solutions that are limited by specified constraints and criteria. Students test for the best solution and improve the solution based on the test results. Upon entering high school, students investigate problems of societal and global importance, use quantitative methods and/or computer simulations to compare and test possible solutions, and improve the solution based on test results. Additionally, high school students consider the global and societal ram-

 DOI: 10.4324/9781003234951-13

ifications of their solutions. Students who are gifted and advanced in STEM do not necessarily need to follow the recommended progressions (Adams, Cotabish, & Ricci, 2014). Gifted and advanced learners, including those from diverse backgrounds, thrive with opportunities to investigate relevant problems of social and global significance (Lovecky, 1994). They should be presented with criteria and constraints early on to increase their opportunities for critical and creative thinking. Furthermore, to increase interest and engagement, gifted and advanced learners need opportunities to improve their solutions, while considering global and societal ramifications, even at the earliest grade levels.

DON'T FORGET THE CONTENT

Many teachers in gifted education classrooms commonly incorporate engineering practices into their curriculum. For example, gifted classrooms often engage in projects such as building towers or bridges, designing protective coverings for egg drop experiments, and creating houses using available 3-D software. These activities are great but can be better if they are embedded in core content (Dailey & Cotabish, 2015). For example, when challenging students to build the tallest tower, teachers might encourage students to consider the reasons behind the tower stability through the examination of Newton's Third Law and balanced and unbalanced forces. In designing the tower, students could examine geometric shapes to determine what will lead to the strongest and most stable tower. Teachers might also incorporate real-world economics into the activity to further engage students in relevant learning.

Many times we think engineering is most appropriate with the study of mathematics or science—in particular, physical science—however, any content area or domain can be used to engage students in the engineering design process. Using fiction or nonfiction stories in literature, students can identify problems and seek solutions. In the early grades, fairy tales are an engaging way to introduce students to the engineering design process. In the fictional world of fairy tales, there are many problems and many opportunities to problem solve. For example, using "Goldilocks and the Three Bears," learners identify a problem with the temperature of the porridge. Learners are then directed to design a device that keeps the porridge "just right." This lesson would be especially appropriate for second-grade students because the study of fairy tales is prominent in the literature strand of the Common Core State Standards for English Language Arts (CCSS-ELA; National Governors Association Center for Best Practices & Council of Chief State School Officers, 2010) for grade 2. Additionally, this lesson would address NGSS grade

2 standards: Matter and Its Interactions. To increase the depth and complexity of the lesson, teachers can challenge advanced learners to explore the kinetic energy of the molecules in the porridge as it is heated and cooled. Advanced students may also be asked to find their own problem in the story and use the engineering design process to develop solutions to the problem.

Engineering also works in a social studies lesson. In the study of economics, students could be challenged to design and create a new coin that is cost effective and durable. Students would explore types of metals that are the least reactive and resistant to corrosion— leading into a study of chemical and physical properties and reactions. Additionally, students could consider aesthetic qualities of the coin, focusing on designing a coin that is both practical and attractive. Through the challenge of designing a new coin, there are myriad other curricular options appropriate across content areas.

The options for integrating engineering into the curriculum are countless and rewarding for both students and teachers. As an example, we provide a scenario below of a fictional teacher trying to meet the NGSS standards. We walk you through the steps our fictional teacher takes as she seeks to increase the lesson's authenticity by adding engineering design processes. Additionally, the lesson plan provides extensions or areas of differentiation for gifted and advanced learners.

Classroom Scenario

Miss Smith's second-grade classroom will begin a plant unit, and she is focusing the unit on what plants need to grow and live. Table 9.1 provides the lesson plan that Miss Smith used in previous classes. The objectives for the lesson identify SWBAT, which stands for "Students will be able to."

After analyzing her previous lesson, Miss Smith was satisfied that the lesson had many good components. The students applied their understanding to a real-world question to summarize the lesson. Students compared and contrasted the growth and development of different types of seeds, and the plant maze provided growth data that students used to show the importance of light for plant growth. Miss Smith also noted that her students critically considered how the maze modeled natural obstructions to light to affect plants' direction of growth. The students communicated their understandings in several ways including journal writing, drawings, and measurement. Furthermore, the advanced student assessment and extension activity provided students a chance to creatively demonstrate their knowledge. An example of what this might look like is available through Blendspace (Buchanan, n.d.). The example uses the aloe plant. A student can create a digital multimedia lesson on how the aloe plant adapted the ability to store water in their enlarged, fleshy leaves.

DESIGNING ENGINEERING CURRICULUM AND ASSESSING STUDENT PERFORMANCE

TABLE 9.1
Miss Smith's Initial Plant Lesson Plan

Teacher Information:
Plants need air, water, nutrients, and light. Students investigate the importance of light for plants to grow. Students learn about the life cycle of a plant.

	Typical Learner(s)	Advanced Learner(s)
Grade 2 **2-LS2-A Interdependent Relationships in Ecosystems:** Plants depend on water and light to grow. (2-LS2-1)		
Performance Expectations		
SWBAT: ○ identify and describe the different stages of a plant's life cycle. ○ compare and contrast monocot and dicot seeds. ○ analyze data, observable and measurable, and use this data to explain the importance of light for plant growth.	Directions: 1. Students observe wet and dry monocot and dicot seeds. 2. Students plant wet seeds in a cup with soil and allow the seeds to sprout. This may take several days, and the students explain observations and measure each day. 3. Students use a light maze (teacher created using a shoebox and extra cardboard) for a bean plant to grow through to represent phototropism. The plant grows and maneuvers through the maze as it responds to the light source.	Directions: 1. Students observe wet and dry monocot and dicot seeds. 2. Students plant wet seeds in a cup with soil and allow the seed to sprout. This may take several days, and the students explain observations and measure each day. 3. Students design and build a light maze using a shoebox, extra cardboard and packing tape for a bean plant to grow through to represent phototropism. The plant grows and maneuvers through the maze as it responds to the light source.
	To **assess** student understanding, students journal the progression of the plant with weekly drawings and measurements. Students estimate the height of the plants	To **assess** student understanding, students journal the progression of the plant with weekly drawings and measurements. Students create a line graph to chart growth

TABLE 9.1, *CONTINUED*

Performance Expectations	Typical Learner(s)	Advanced Learner(s)
	(once they sprout) in lengths using units of inches and centimeters. At the end of each week, students determine how much taller the plant has grown by showing this difference in inches and centimeters.	progression. At the end of each week, students determine how much taller the plant has grown by showing this difference in inches and centimeters. They discuss the actions/movement of the plant's growth. Students also discuss how plants grow in different habitats (extending to LS.4.D, 2-LS4-1).
	Assessment Question: How does this experiment model the way plants might grow in nature?	**Assessment Challenge:** Using this experiment as a model, create a scenario that explains the way plants might grow in nature.
	For further challenge, students use Blendspace to create a photo story on the life cycle of a plant.	For further challenge, advanced students use Blendspace to demonstrate student's knowledge of how plants grow in different habitats.
Implementation	**Materials:** Shoe box with lid and hole cut out of one end, bean seeds, corn seeds, cup, potting soil, extra cardboard, packing tape, and a variety of materials for maze.	

Miss Smith's recent professional development experiences focused on science and engineering practices, and she noticed that her lesson did not meet all of the needed NGSS. For her students to meet 2-LS2-1, *Plan and conduct an investigation to determine if plants need sunlight and water to grow*, they must plan and conduct their own investigations. Although her original lesson provided students valuable content and skills, Miss Smith noted the students were not defining problems with specific criteria for multiple solutions. There were no testing prototypes or refining and optimizing solutions in the lessons (engineering design). The students' assessment did apply their learning to a real-world question, but the set of lessons, as a whole, lacked authenticity.

She begins with thinking about how to give her students more *voice* and *choice* with this lesson. Student choice provides a pathway for students to partake in the design of the learning process, and this participation allows students to find meaning in the content through their own means. Students can choose the type of plant, the amount of sunlight and water the plant will receive, and the container to hold the plant. This brainstorming allows Miss Smith to develop an essential question: *What do plants need in order to live and grow?* An essential question sets the stage for students to explore the content information related to the problem. After this research, students develop various strategies or action plans to generate a solution. A good essential question (Global Digital Citizen Foundation, 2016):

- inspires a quest for knowledge and discovery,
- encourages and develops critical thinking processes,
- leads students to engineer real-world solutions for real-world problems, and
- is all about possibilities. (para. 2)

When answering an essential question, students need opportunities to continue the investigation by asking more questions for multiple design solutions—Who, what, when, where, why, and how? After asking these questions, students can further research information that is pertinent to the essential question.

Miss Smith considers the NGSS standard 2-LS2-1, *Plan and conduct an investigation to determine if plants need sunlight and water to grow*, and evaluates the essential question:

- Do her students need to question and research information in order to answer the essential question? Yes.
- Does the essential question require multiple investigations to observe and question, organize, and analyze data in order to develop a comprehensive summary? Yes.
- Does the essential question incorporate real-world applications? Yes.
- Does the essential question have a definite answer? No.

With the essential question identified, she prepares to develop the lesson through the engineering design process. According to the NGSS (NGSS Lead States, 2013), the core idea of engineering design includes three component ideas:

- Defining and delimiting engineering problems involves stating the problem to be solved as clearly as possible in terms of criteria for success, and constraints or limits.
- Designing solutions to engineering problems begins with generating a number of different possible solutions, then evaluating potential solutions to see which one best meets the criteria and constraints of the problem.
- Optimizing the design solution involves a process in which solutions are systematically tested and refined and the final design is improved by trading off less important features for those that are more important. (p. 104)

Table 9.2 provides the updated lesson plan after Miss Smith integrated engineering design processes to increase the authenticity of the lesson. She created an anchor chart to include information about the engineering design process that includes brainstorming solutions to a problem, planning an investigation based off their brainstorming, implementing the plan, and evaluating the solutions to determine how well they meet the challenge. Her students wrote these steps in their journals to help them prepare for the challenge.

The following scenario was provided:

A local neighborhood wants to plant a garden in a corner of the community playground. In the past they have tried to plant the garden, but the plants did not grow and produce fruits and vegetables. When it rains, the water is not absorbed into the soil. It either runs off quickly or floods the area of the park. The local gardener thinks something needs to change so that water can be trapped and then absorbed by the plants. What do you need to know so that you can design a container that would hold the appropriate amount of water to nourish a bean seed? Once you have designed, tested, and revised the container, communicate your findings to the local gardener.

DESIGNING ENGINEERING CURRICULUM AND ASSESSING STUDENT PERFORMANCE

TABLE 9.2

Miss Smith's Modified Lesson Plan With Engineering Design Processes

Grade 2

2-LS2-A Interdependent Relationships in Ecosystems: Plants depend on water and light to grow. (2-LS2-1)

Teacher Information:

Plants need air, water, nutrients, and light. Students investigate the importance of light for plants to grow. Students learn about the life cycle of a plant. Students examine the effects of different amounts of light and water on green plants, introducing the processes of photosynthesis and transpiration. Students make predictions and record observations.

Performance Expectations	Typical Learner(s)	Advanced Learner(s)
SWBAT: ○ identify and describe the different stages of a plant's life cycle. ○ compare and contrast monocot and dicot seeds. ○ analyze data, observable and measurable, and use this data to explain the importance of light for plant growth.	**Essential Question:** What do plants need in order to live and grow? **Part A Directions:** 1. Students are presented the scenario problem and discuss possible solutions. 2. After these discussions, students begin to create a KLEWS chart, either using a class anchor chart or in their journals. KLEWS: What do we think we KNOW? What are we LEARNING? What is our EVIDENCE? What are we WONDERING? What SCIENTIFIC principles/vocabulary help explain the phenomena? 3. The students begin with the previous experiments. 4. Students plant wet seeds in a cup with soil and allow the seeds to sprout. This may take several days, and the students explain observations and measure each day.	**Essential Question:** What do plants need in order to live and grow? **Part A Directions:** 1. Students are presented the scenario problem and discuss possible solutions. For advanced students, instead of focusing on the container, they will focus on the "best" combination of soils to hold the appropriate amount of water to nourish a bean seed. 2. After these discussions, students begin to create a KLEWS chart either using a class anchor chart or in their journals. KLEWS: What do we think we KNOW? What are we LEARNING? What is our EVIDENCE? What are we WONDERING? What SCIENTIFIC principles/vocabulary help explain the phenomena? 3. Students observe wet and dry monocot and dicot seeds.

TABLE 9.2, *CONTINUED*

Performance Expectations	Typical Learner(s)	Advanced Learner(s)
	5. After these comparisons, students revisit the KLEWS chart. *To help students make meaning of what they are learning, the teacher facilitates reasoning involved in the LE and S sections of the chart.* 6. Next, the student groups plant beans or corn seeds for further light and water investigations. For the seed of their choice, each student group will investigate different variables: light versus dark, and watered versus nonwatered conditions (students will test one variable at a time). 7. Students will predict and measure the growth of the seedlings using nonstandard measurement throughout these investigations using everyday objects such as paper clips or number cubes to measure plant height. They will count number of leaves and sketch the root structure. Finally, each group will share the results from the tests.	8. Students plant wet seeds in a cup with soil and allow the seeds to sprout. This may take several days, and the students explain observations and measure each day. 9. After these comparisons, students revisit the KLEWS chart. *To help students make meaning of what they are learning, the teacher facilitates reasoning involved in the LE and S sections of the chart.* 10. Next, the student groups plant beans and corn seeds for further light and water investigations. For the seed of their choice, each student group will investigate different variables: light versus dark, and watered versus nonwatered conditions. Multiple tests can be used on that one variable. 11. Students will predict and measure the growth of the seedlings using nonstandard measurement throughout these investigations using everyday objects such as paper clips or number cubes to measure plant height. They will

DESIGNING ENGINEERING CURRICULUM AND ASSESSING STUDENT PERFORMANCE

TABLE 9.2, *CONTINUED*

Performance Expectations	Typical Learner(s)	Advanced Learner(s)
		count number of leaves and sketch the root structure. 12. Finally, each group will share the results from the tests. Students will then compare and contrast the data from the different conditions using pictorial bar graphs.
	To **assess** student understanding, students journal the progression of the plant with weekly drawings and measurements. Students can estimate the height of the plants (once they sprout) in lengths using units of inches and centimeters. At the end of each week, students can determine how much taller the plant has grown by showing this difference in inches and centimeters. Students should continuously revisit the KLEWS chart.	To **assess** student understanding, students journal the progression of the plant with weekly drawings and measurements. Students create a line graph to chart plant growth progression. At the end of each week, students can determine how much taller the plant has grown by showing this difference in inches and centimeters. Students should continuously revisit the KLEWS chart.
	Assessment Question: What are the optimal environments for growing crops and other plants used to produce food or crops?	**Assessment Question:** What is needed for growing crops and other plants used to produce products in different areas with different soil and water conditions?

TABLE 9.2, *CONTINUED*

Performance Expectations	Typical Learner(s)	Advanced Learner(s)
	Part B Directions: 1. Students **discuss** what they know and need to know to answer Part A's Assessment Question: What are the optimal environments for growing crops and other plants used to produce products? 2. Student groups use the information learned from Part A to **brainstorm** possible answers to the Assessment Question. 3. After developing several possible solutions, the student groups **plan** to test a chosen set of optimal conditions for a bean seed or corn seed to live and grow. Each group **presents** the optimal conditions tested and the results (plant height, number of leaves, root structure). 4. After the presentations, student groups discuss the results of every presentation. Together they **reimagine** the optimal conditions needed and retest with the change(s) identified as needed. 5. Student groups **present,** their findings from this retesting. 6. Students should continuously revisit the KLEWS chart.	**Part B Directions:** 1. Student groups are given an area of the world that is struggling with crop development. Each group researches a specific area to test. 2. Students **discuss** what they know and need to know to answer Part A's Assessment Question: What are the optimal environments for growing crops and other plants used to produce products? Student groups use the information learned from Part A to **brainstorm** possible answers to the Assessment Question. 3. After developing several possible solutions, the student groups **plan** to test a chosen set of optimal conditions for a bean seed or corn seed to live and grow. Each group **presents** the optimal conditions tested and the results (plant height, number of leaves, root structure). 4. After the presentations, student groups discuss the results of every presentation. Together they **reimagine** the optimal conditions needed and retest with the change(s) identified as needed.

DESIGNING ENGINEERING CURRICULUM AND ASSESSING STUDENT PERFORMANCE

TABLE 9.2, *CONTINUED*

Performance Expectations	Typical Learner(s)	Advanced Learner(s)
		5. Student groups **present**, their findings from this retesting. 6. Students should continuously revisit the KLEWS chart.
	To **assess** student understanding, students journal the progression of the plant with weekly drawings and measurements. Students can estimate the height of the plants (once they sprout) in lengths using units of inches and centimeters. At the end of each week, students can determine how much taller the plant has grown by showing this difference in inches and centimeters. For further challenge, students could use Blendspace to create a photo story on the life cycle of a plant. An extended project could challenge the students to design and create a school garden with the optimal sunlight needed for specified types of plants.	To **assess** student understanding, students journal the progression of the plant with weekly drawings and measurements. Students can estimate the height of the plants (once they sprout) in lengths using units of inches and centimeters. At the end of each week, students can determine how much taller the plant has grown by showing this difference in inches and centimeters. Students can also create a line graph. For further challenge, advanced students could use Blendspace to demonstrate student's knowledge of how plants grow in different habitats (extending to LS.4.D, 2-LS4-1). An extended project could challenge the students to design and create a school garden with the optimal sunlight needed for various types of plants.
	Part C Directions: Students **discuss** what they know and need to know to solve the local gardener's	**Part C Directions:** Students **discuss** what they know and need to know to solve the local gardener's problem:

TABLE 9.2, *CONTINUED*

Performance Expectations	Typical Learner(s)	Advanced Learner(s)
	problem: Design a container that would hold the appropriate amount of water to nourish a bean seed. "From your testing you found that if you watered your bean seed too much the seed would either float or even grow mold from being too wet. You found if you poked holes in the cup, much of the water would leak out. With this in mind, design a container that would hold the appropriate amount of water to nourish the seed. Once you have designed, tested, and revised your findings to the local gardener."	Design the best mix of soil that would be optimal for moisture control to nourish a bean seed. "From your testing you found that if you watered your bean seed too much the seed would either float or even grow mold from being too wet. With this in mind, consider different soils and design the best mix of soil that would be optimal for moisture control to nourish the seed. Once you have designed, tested, and revised your plan for the best mix of soil, communicate your findings to the local gardener."
	To assess student understanding, students journal the steps taken during the creation of the container to show strengths and weaknesses of each design. Students should continuously revisit the KLEWS chart.	To assess student understanding, students journal the steps taken during the mixing of the soil types to show strengths and weaknesses of each combination. Students should continuously revisit the KLEWS chart.
	For further challenge, students could use Blendspace to create a photo story evidencing their steps taken to solve the problem.	For further challenge, students could use Blendspace to create a photo story evidencing their steps taken to solve the problem.
Implementation	**Materials:** Bean seeds, corn seeds, cup, potting soil, various objects to make containers for the bean plants. Several large plastic containers for students to submerge their created containers and test their effectiveness. For advanced students who test different soil combinations, provide soil of various amounts of sand and clay.	

Miss Smith's updated lesson offers her students the opportunity to develop several possible solutions; the students design the optimal conditions for a bean seed and/or corn seed to live and grow. The essential question provides multiple, real-world application opportunities. Students can investigate the growth of local plants or be challenged with a scenario to improve crop production for a third-world country. Adding the Know, Learning, Evidence, Wonder, Scientific Principles/Vocabulary (KLEWS) chart helped the students make meaning of the information throughout the solving of the problem (Hershberger, Zembal-Saul, & Starr, 2005). "The L and E of KLEWS are aligned with claims and evidence, while the S is related to the scientific reasoning needed to build a complete explanation" (Hershberger & Zembal-Saul, 2015, p. 68). Miss Smith's students were able to connect their prior knowledge of what plants needed to grow to the evidence they were collecting, and then they were able to apply new vocabulary and scientific principles to their observations. This either affirmed what they knew or allowed new connections to be made. By allowing the KLEWS chart to be completed in any order (not in order of left to right), Miss Smith's students interacted with the KLEWS chart as they collected evidence and constructed meaning. The updated lesson allows students to practice problem-solving skills through the engineering design process. Asking about what they know and need to know about a problem is a metacognitive skill students of all ages should practice. Students brainstorm solutions to a problem, plan an investigation based off their brainstorming, implement the plan, and evaluate the solutions to determine how well they meet the challenge.

The engineering design process adds the process skills of reimagining and redesigning. Although standard demonstration activities control these process skills, inquiry investigations put the control back onto the students to construct a reasonable explanation. Physical sciences are often thought of when creating engineering lessons. The engineering design process is not just about constructing a product, it is about practicing these process skills—skills that will be used in the real world of scientists or engineers. Furthermore, as demonstrated in the scenario above, content was the key ingredient. Inquiry and engineering activities without content (*hands on* without *minds on*) does not equate to cognitive engagement leading to student understanding and conceptual learning (Martin, 2012).

CONCLUSIONS AND KEY CONSIDERATIONS

Integrating engineering design processes into existing curriculum should not be overwhelming for teachers. One place to start is with the NGSS standards.

Milano (2013) and Dailey and Cotabish (2015) provided examples of how to bundle NGSS performance expectations with instructional questions to guide student learning and lead them through the engineering design process (see Table 9.3). This type of seamless organization helps ensure the engineering design processes are embedded in content for relevant and meaningful learning.

Additionally, integrating engineering design processes in the curriculum is beneficial to advanced and gifted learners. Gifted learners benefit from engaging in real-world problem solving, reflection and collaboration, and creative and critical thinking (Robbins, 2011; VanTassel-Baska, 1998)—typical instructional pieces found in engineering design processes. Furthermore, as demonstrated in Tables 9.1–9.2, teachers can differentiate these types of learning experiences by providing greater complexity and more constraints to the engineering problems.

DESIGNING ENGINEERING CURRICULUM AND ASSESSING STUDENT PERFORMANCE

TABLE 9.3
Energy, Energy Transfer, and Engineering Design Bundle

	Energy, Energy Transfer, and Engineering Design Bundle	Instructional Questions
MS-PS3-3.	**Apply scientific principles to design, construct, and test a device that either minimizes or maximizes thermal energy transfer.** [Clarification Statement: Examples of devices could include an insulated box, a solar cooker, and a Styrofoam cup.] [Assessment Boundary: Assessment does not include calculating the total amount of thermal energy transferred.]	What kind of device would keep your cup of coffee warm during a cold football game?
MS-PS3-4.	**Plan an investigation to determine the relationships among the energy transferred, the type of matter, the mass, and the change in the average kinetic energy of the particles as measured by the temperature of the sample.** [Clarification Statement: Examples of experiments could include comparing final water temperatures after different masses of ice melted in the same volume of water with the same initial temperature, the temperature change of samples of different materials with the same mass as they cool or heat in the environment, or the same material with different masses when a specific amount of energy is added.] [Assessment Boundary: Assessment does not include calculating the total amount of thermal energy transferred.]	How does the kinetic energy of your coffee compare in the devices that you designed and created?
MS-PS3-5.	**Construct, use, and present arguments to support the claim that when the kinetic energy of an object changes, energy is transferred to or from the object.** [Clarification Statement: Examples of empirical evidence used in arguments could include an inventory or other representation of the energy before and after the transfer in the form of temperature changes or motion of object.] [Assessment Boundary: Assessment does not include calculations of energy.]	What happened to the kinetic energy of the molecules in the coffee when the coffee was allowed to cool? What evidence do you have to support your claim? (Where did the heat energy go?)
MS-ETS1-4.	Develop a model to generate data for iterative testing and modification of a proposed object, tool, or process such that an optimal design can be achieved.	How can your device be improved?

Adapted from "The Next Generation Science Standards and Engineering for Young Learners: Beyond Bridges and Egg Drops," by M. Milano, 2013, *Science and Children*, 51(2), 10–16. Copyright 2013 by National Science Teachers Association. Adapted with permission.

REFERENCES

Adams, C., Cotabish, A., & Ricci, M. C. (2014). *Using the Next Generation Science Standards with gifted and advanced learners.* Waco, TX: Prufrock Press.

Buchanan, M. (n.d.). Why plants grow the way they do. *Blendspace.* Retrieved from https://www.tes.com/lessons/GYT2vC_xEFhZIQ/why-plants-grow-the-way-they-do

Dailey, D., & Cotabish, A. (2015). Implementing engineering practices with advanced learners. In B. MacFarlane (Ed.), *STEM education for high-ability learners: Designing and implementing programing* (pp. 71–84). Waco, TX: Prufrock Press.

Global Digital Citizen Foundation. (2016). *Writing essential questions.* Retrieved from http://help.solutionfluency.com/article/49-writing-essential-questions

Hershberger, K., Zembal-Saul, C., & Starr, M. L. (2005). Evidence helps KWL get a KLEW. *Science and Children, 43*(5), 50–53.

Hershberger, K., & Zembal-Saul, C. (2015). KLEWS to explanation-building in science: An update to the KLEW chart adds a tool for explanation building. *Science and Children, 52*(6), 66–71.

Lovecky, D. V. (1994). Exceptionally gifted children: Different minds. *Roeper Review, 17,* 116–120.

Martin, D. J. (2012). *Elementary science methods: A constructivist approach* (6th ed., pp. 2–14). Belmont, CA: Wadsworth.

Milano, M. (2013). The Next Generation Science Standards and engineering for young learners: Beyond bridges and egg drops. *Science and Children, 51*(2), 10–16.

National Governors Association Center for Best Practices, & Council of Chief State School Officers. (2010). *Common Core State Standards for English language arts.* Washington, DC: Author.

NGSS Lead States. (2013). *Next generation science standards: For states, by states.* Washington, DC: The National Academies Press.

Robbins, J. I. (2011). Adapting science curricula for high-ability learners. In J. VanTassel-Baska & C. A. Little (Eds.), *Content-based curriculum for high-ability learners* (2nd ed., pp. 217–238). Waco, TX: Prufrock Press.

VanTassel-Baska, J. (1998). Planning science programs for high ability learners. *ERIC Clearinghouse on Disabilities and Gifted Education.* Retrieved from http://eric.ed.gov/?id=ED425567

CHAPTER 10

Integrating Problem-Based Learning Into Engineering Curriculum for High-Ability Learners

Joyce VanTassel-Baska and Bronwyn MacFarlane

These kids collect and then create. They love to make geometric bottles on paper. They keep interesting things in their pockets, and find interesting things in the trash. Everything is usable and taken apart to see what is inside. They are not good at in-seat behavior nor attending to formal class formats. They are often alone in their behavior with objects, taking apart complex mechanical materials. They use the pencil sharpener because of fascination with how it works. They love zippers and fasteners. One 3-year old taught himself how to zip his coat and put on his shoes. They are also big into water, loving properties such as spillage. One young

 DOI: 10.4324/9781003234951-14

student made a waterfall out of a pitcher and basin to observe large-scale spillage. They love to figure out how things work.

These are the comments of a preschool teacher on behaviors she has observed in a few of her 3- and 4-year-olds—behaviors that presage engineers and scientists and mathematicians of the future, behaviors that begin young and are sustained through interest and passion with materials and processes applied in systematic ways. Research on the abilities in childhood associated with engineering talent later in life including the following (U.S. Department of Labor, n.d.):

- **Problem Sensitivity:** The ability to tell when something is wrong or is likely to go wrong. It does not involve solving the problem, only recognizing that there is a problem.
- **Deductive Reasoning:** The ability to apply general rules to specific problems.
- **Inductive Reasoning:** The ability to combine pieces of information to form general rules or conclusions (includes finding a relationship among seemingly unrelated events).
- **Mathematical Reasoning:** The ability to choose the right mathematical methods or formulas to solve a problem.
- **Number Facility:** The ability to add, subtract, multiply, or divide quickly and correctly.
- **Perceptual Speed:** The ability to quickly and accurately compare similarities and differences among sets of letters, numbers, objects, pictures, or patterns. The things to be compared may be presented at the same time or one after the other. This ability also includes comparing a presented object with a remembered object.
- **Control Precision:** The ability to quickly and repeatedly adjust the controls of a machine or a vehicle to exact positions.

And then the skills associated with future STEM career professionals manifest among gifted students in the following ways (West, 2012):

- **Critical Thinking:** Using logic and reasoning to identify the strengths and weaknesses of alternative solutions, conclusions, or approaches to problems.
- **Active Learning:** Understanding the implications of new information for both current and future problem-solving and decision-making.
- **Complex Problem Solving:** Identifying complex problems and reviewing related information to develop and evaluate options and implement solutions.
- **Operations Analysis:** Analyzing needs and product requirements to create a design.

- **Technology Design:** Generating or adapting equipment and technology to serve user needs.
- **Systems Analysis:** Determining how a system should work and how changes in conditions, operations, and the environment will affect outcomes.
- **Systems Evaluation:** Identifying measures or indicators of system performance and the actions needed to improve or correct performance, relative to the goals of the system. (p. 2)

There is clearly a match between these early observable behaviors and the endgame of increasing the percentage of U.S. citizens in STEM (science, technology, engineering, and math) careers, especially engineering. This chapter explores what we know about best practices in delivering challenging curriculum for the gifted and is designed to stimulate the readers' thinking about developing learning opportunities focused on engineering with the specific use of the instructional model of problem-based learning (PBL), which mirrors the professional process models used in the engineering field and discussed later in the chapter. As the reader reflects upon the details presented herein about what works in PBL curriculum, consider the engineering problems in Table 10.1 and their application to the curriculum models and ideas presented.

THE PAST AND THE FUTURE: HISTORICAL CONTEXT FOR DELIVERING AND IMPROVING STEM EDUCATION

The need for improved STEM education curriculum has been an ongoing concern for some time. While demand for STEM career graduates continues to increase, concern for improving students' STEM skills has also continued to provide a conundrum for educators. Twenty years after Project 2061 was initiated, the 2005 results from the National Assessment of Education Progress (NAEP) indicated that students in all grade levels showed a lack of understanding of scientific concepts and reasoning (Grigg, Lauko, & Brockway, 2006). More than a decade after declaring that the United States will be the first in the world in science, U.S. students scored behind other countries on the Trends in International Mathematics and Science Study (TIMSS; Gonzales et al., 2008) and the Program for International Student Assessment (PISA; OECD, 2007). Further, U.S. students scored lower than 12 other countries/jurisdictions on the PISA when com-

TABLE 10.1
Problem-Based Engineering Learning Opportunities

Problems for engineering scenarios can take a variety of forms for students to analyze. Consider the following problems for students to explore in depth and complexity as they advance in their understanding and skills.
1. To increase two-way communication by building a transistor radio.   2. To access clean water by solving how to transport water to disconnected villages located around the globe.   3. To produce energy using water power through a hydroelectric dam.   4. To harness natural airflow around the planet to produce electrical energy.   5. To mechanize artificial limbs for medical use.   6. To propel an automobile by generating wind energy.   7. To develop webpage code to illustrate a science concept with animation.   8. To write a computer program or create an app for a mobile phone.

paring scores of the highest achieving students on the combined science literacy scale (Baldi, Jin, Green, & Herget, 2007).

Current science curriculum and instruction are failing our highest achieving students. In a similar way, math talent is also languid, with students in the United States consistently ranking somewhere in the middle (Gallagher, 1989). Educators must be compelled by these data to use curriculum and engage in instruction that encourages our most scientifically talented students and all students to develop scientific habits of mind, to investigate natural phenomena, and to become engaged in scientific problem solving if we are to reach and surpass our global competition. For the most able students in science and math, educational goals must transcend minimal standards of scientific literacy.

In the past decade, several national reports have called for increased STEM education, including suggestions for earlier intervention, foci on the most able children, and renewed interest in the importance of spatial ability for STEM innovation. In particular, the National Science Board (2010) details the lack of STEM preparation in schools and outlines an agenda for action in their report, *Preparing the Next Generation of STEM Innovators.* The report notes that while many others have made recommendations focusing on raising overall performance of America's students, few have "focused on raising the ceiling of achievement for our Nation's most talented and motivated students" (p. 4). The National Science Board further outlines key issues, including the importance of early intervention and identification of spatial talent as a specific area for attention. Innovation is the clarion call of the report, which calls for priming the pipeline for scientists, technology specialists, engineers, and mathematicians, who can solve the real-world problems we face, as well as provide a competitive edge.

THE NEED FOR RIGOROUS CURRICULUM TO NURTURE SCIENCE TALENT ACROSS K-12 EDUCATION

Engineering is the applied field of science and mathematics, and students need early exposure and interaction with related ideas from these fields to pursue engineering as a career. Data from longitudinal early childhood studies have demonstrated that development and implementation of intensive, high-quality, pervasive interventions can impact achievement patterns early for low-income learners (Borman & Hewes, 2002; Ramey & Ramey, 1998). Further, research in gifted education has shown that low-income gifted learners can benefit from targeted interventions in content areas that are focused on higher level skill development within specific content areas, including science (Feng, VanTassel-Baska, Quek, Bai, & O'Neill, 2005; Gavin et al., 2007; VanTassel-Baska, Avery, Hughes, & Little, 2000; VanTassel-Baska, Bass, Ries, Poland, & Avery, 1998; VanTassel-Baska, Zuo, Avery, & Little, 2002). Research from the Advanced Placement (AP) and International Baccalaureate (IB) programs has shown that low-income learners who do not have prerequisite skills (such as writing, study skills, and time management) necessary for success in these courses are unable to acquire the skills fast enough within the courses to be successful (Hertberg-Davis & Callahan, 2008). Furthermore, longitudinal research from the Study of Mathematically Precocious Youth (SMPY) has revealed that a confluence of factors contributes to the development of scientific expertise over the lifespan, including investigative interests and a focus on finding truth through cognitive means, ability in math, high levels of spatial ability, a sustained commitment to scientific pursuits, dedication to school and work within and outside the school/work environment, and specific educational opportunities (Lubinski & Benbow, 2006). For all learners, including those from low-income backgrounds, educational opportunities in the form of rigorous science curriculum and instruction scaffolded with prerequisite skills can be helpful to teach science literacy, develop the scientific habits of mind necessary for success, and nurture scientific talent, so necessary for engineering success.

Our own studies from Project Clarion at William & Mary suggest that early intervention in the scientific concepts and skills, promoted in the new science standards as well as in the emphases in engineering programs grades K–12, produce significant gains when compared to matched groups of students. These studies have been conducted with students in grades 1–3. Further analyses suggest that the younger students grew the most across a 2-year period, providing further evidence that starting earlier has immediate benefits (Kim et al., 2012). This study also suggests that science curriculum, designed according to exemplary science strategies and differentiation approaches for high-ability learners, may be used

DESIGNING ENGINEERING CURRICULUM AND ASSESSING STUDENT PERFORMANCE

successfully with all learners to enhance critical thinking abilities in general, as well as to build science skills and concepts. What we have learned about science curriculum is transferrable to developing engineering curriculum.

ENGINEERING EDUCATION CURRICULUM

Engineering has long been considered the field of expertise leading to the solutions of problems that have otherwise been hindered by various obstacles. Engineers respond to challenges posed by both the natural and physical world. These problems are complex and span healthcare, technology, energy, manufacturing and production, transportation, the environment, and more. However, engineering education seldom addresses these complex, multifaceted, and challenging issues facing engineers and instead has been constrained through schooling for the following reasons: (1) Engineering lacks definition for the general public and is viewed broadly; (2) for children in elementary school, engineering is viewed in terms of construction for building small machines and structures; and (3) when these students enter secondary school, they view engineering as a high-level, mathematically dense field related to constructing bridges and spacecraft. These limited views have led to an incomplete and inaccurate understanding of engineering and subsequent lack of interest in post secondary engineering studies (Lawrence, Hinterlong, & Sutherland, 2015, p. 392).

Just as gifted education programs should define which gifts and talents will be identified and cultivated in gifted services, so too, should STEM educational initiatives define STEM and how STEM-related topics will be operationalized in a precollegiate curriculum program (MacFarlane, 2016, p. 144). The importance of operationalizing a well-defined, planned set of learning goals as outlined by MacFarlane (2016) is not to be undervalued. Engineering education has suffered in K–12 as a result of society not clearly understanding what engineering is, which makes it all the more imperative that any educational engineering initiatives for the gifted clarify relevant student learning outcomes as a result of program participation.

A Framework for K–12 Science Education (National Research Council [NRC], 2012) provides the blueprint for the Next Generation Science Standards (NGSS) and expresses a vision in science education that requires students to operate at the nexus of three dimensions of learning: science and engineering practices, crosscutting concepts, and disciplinary core ideas. The eight practices of science and engineering that the framework (NRC, 2012) identifies as essential for all students to learn are:

1. asking (for science) questions and defining problems (for engineering),
2. developing and using models,
3. planning and carrying out investigations,
4. analyzing and interpreting data,
5. using mathematics and computational thinking,
6. constructing explanations (for science) and designing solutions (for engineering),
7. engaging in argument from evidence, and
8. obtaining, evaluating, and communicating information. (p. 3)

The integrated nature of engineering as a discipline supports the applied use of science and mathematics. In addition to the engineering design guidelines, thoughtful integration of the standards—the Common Core State Standards for Mathematics (CCSSM), the CCSS for English Language Arts (CCSS-ELA), NGSS, and 21st-century skills—will enhance the applied understandings developed in a well-defined engineering curriculum. Let's take a look at the NGSS engineering design standards at the middle and high school levels (NGSS Lead States, 2013) and consider curriculum alignment. The progressive knowledge and skill understanding that builds between middle and high school is well illustrated between Tables 10.2 and 10.3. Alignment of the curriculum to these NGSS standards can increase the robustness of the learning activities designed for students.

APPLIED ENGINEERING:
THE 4-H PROJECT LEARNING MODEL

Many educational resources are available for those charged with beginning and developing engineering curriculum for special programs. 4-H, the nation's largest informal youth education program, made a commitment to help address youth science literacy needs with a variety of STEM-related project curriculum materials. A 2013 report about the impact of 4-H project involvement found that 4-H youth are 2 times as likely to participate in afterschool science, engineering, or computer technology programs (Positive Development of Youth, 2013). 4-H STEM programs combine the strengths of experiential, hands-on education and inquiry-based science learning with a positive youth development framework that holistically addresses the developmental and educational needs of young people. Through participation in 4-H engineering and technology projects, youths develop technological literacy and learn about the iterative engineering design

TABLE 10.2
Next Generation Science Standards for
Engineering Design–Middle School

Students who demonstrate understanding can:	
MS-ETS1-1.	Define the criteria and constraints of a design problem with sufficient precision to ensure a successful solution, taking into account relevant scientific principles and potential impacts on people and the natural environment that may limit possible solutions.
MS-ETS1-2.	Evaluate competing design solutions using a systematic process to determine how well they meet the criteria and constraints of the problem.
MS-ETS1-3.	Analyze data from tests to determine similarities and differences among several design solutions to identify the best characteristics of each that can be combined into a new solution to better meet the criteria for success.
MS-ETS1-4.	Develop a model to generate data for iterative testing and modification of a proposed object, tool, or process such that an optimal design can be achieved

TABLE 10.3
Next Generation Science Standards for
Engineering Design–High School

Students who demonstrate understanding can:	
HS-ETS1-1.	Analyze a major global challenge to specify qualitative and quantitative criteria and constraints for solutions that account for societal needs and wants.
HS-ETS1-2.	Design a solution to a complex, real-world problem by breaking it down into smaller, more manageable problems that can be solved through engineering.
HS-ETS1-3.	Evaluate a solution to a complex, real-world problem based on prioritized criteria and trade-offs that account for a range of constraints, including cost, safety, reliability, and aesthetics as well as possible social, cultural, and environmental impacts.
HS-ETS1-4.	Use a computer simulation to model the impact of proposed solutions to a complex, real-world problem with numerous criteria and constraints on interactions within and between systems relevant to the problem.

process, attributes of design, impacts of systems, and effects of technology on the environment.

The learning goals associated with relevant STEM projects are listed below and specifically indicate that 4-H youth will (Texas A&M AgriLife Extension Service, 2015):

- develop an understanding of basic science concepts related to robotics;
- apply the processes of scientific inquiry and engineering design;
- build skills in science, engineering, and technology;
- use the tools of technology to enhance their learning;
- explore related careers in these fields; and
- apply the skills and knowledge they are developing to new challenges. (para. 2)

Let's look closer at the tiered nature of the 4-H robotics curriculum project to see how this engineering curriculum has been constructed for students. The 4-H Robotics Curriculum is comprised of three tracks; each is designed to meet the diverse requirements of 4-H clubs, afterschool programs, and individual youth and school enrichment activities. The curriculum was developed for all levels of expertise—from beginning to advanced. 4-H curriculum can be compacted and accelerated for gifted children to advance with expert project leaders through the project curriculum.

In the project, "4-H Robotics: Engineering for Today and Tomorrow," the curriculum is organized in three tracks to develop youths' skills from beginning to advanced levels. In Track 1 (Virtual Robotics), students use virtual activities online or via DVD-ROM to build a knowledge base about robotic functions and specifically robot arms. In Track 2 (Junk Drawer Robotics), students are encouraged to use household materials to engineer robotic designs in order to accomplish a task. In Track 3 (Platforms), students with access to specific educational robotics platforms (LEGO, VEX, Robotix, etc.) focus on technology and building their engineered plans to work.

The 4-H curriculum further outlines the tiered sequencing of building understanding about robotics through three levels. Each level has three distinct ideas: to learn, to do, and to make. In the robotics tiered project, 4-H'ers first learn about robot arms and designs that allow the robot to grab, lift, and move. Then, at the next level, the mobility of the robots is explored with motors, gears, and buoyancy. Finally, electronic devices, sensors, and programming elements are explored in the third level (California 4-H Youth Development Program, 2016). Much can be learned from the more than a century of 4-H project curriculum offering applied science learning opportunities to children. At each level, 4-H'ers learn about the use of robots in the world, increasingly explore and engage in engineering design processes with appropriate challenges, and experience scientific inquiry. Next, let's consider the instructional elements of school-based methods.

INSTRUCTIONAL FEATURES OF RIGOROUS SCIENCE CURRICULUM

Inquiry-based instructional approaches have traditionally been found to be effective in teaching science. For example, problem-based science curriculum and project-based learning have been found to be effective with gifted and high-ability students who have shown gains in achievement on knowledge acquisition, knowledge application, and science investigation skills at the primary, intermediate, middle, or high school levels (Mioduser & Betzer, 2008; VanTassel-Baska et al., 1998). Further, Swanson (2006) found that problem-based curriculum also produced gains in science achievement with gifted students from low-income backgrounds. Rayneri, Gerber, and Wiley (2006) found that gifted students show preferences for hands-on learning in science. Science curriculum that integrates high-level content, scientific processes, and authentic products, and is concept based has been found to significantly enhance low-income primary and elementary gifted students' achievement in science (Feng et al., 2005; Kim et al., 2012). Finally, VanTassel-Baska, Feng, and Brown (2008) found that implementing well-designed research-based curriculum units differentiated for gifted learners improved teachers' general use of differentiation strategies after receiving consistent professional development in the differentiated instructional strategies.

Targeted interventions that include well-designed, rigorous curriculum provided to low-income gifted learners early in their schooling are critical for such learners' future success. The Integrated Curriculum Model (ICM) was used as the basis to develop the science curriculum in the Project Clarion study (Kim et al., 2012) as well as other studies on early intervention success in science (see Robinson, Dailey, Hughes, & Cotabish, 2014). ICM provided the framework for the integration of content knowledge with concept development and higher level scientific research processes, differentiation strategies, and scaffolding to meet the needs of all learners.

THE CURRICULUM FRAMEWORK FOR WORKING WITH GIFTED LEARNERS IN STEM

Planning curriculum should begin with the design of a curriculum framework that organizes curricula across all levels of schooling within and beyond subject areas. The documents that represent this macroplanning effort include a curriculum framework that articulates goals, outcomes, strategies, and activities across grades K–12. Common goals for a school's STEM curriculum framework for the gifted might include:

- to develop critical thinking skills;
- to develop creative thinking and problem-solving skills;
- to understand the macroconcepts of patterns, change, and systems;
- to develop advanced content skills in all relevant areas of learning; and
- to develop self-understanding and understanding of others.

Each of these goals would be further developed by delineating outcomes, strategies, and assessments that would flesh out the blueprint for further development at each grade level and in each relevant content area of math and science.

Recommended goals for a STEM curriculum might look like the following:

- to develop an understanding of scientific concepts;
- to develop STEM-based skills and abilities;
- to develop an understanding of the scientific research process, including its design and implementation;
- to develop scientific habits of mind (e.g., objectivity, skepticism); and
- to provide mentorship and internship opportunities;

Overlap emphases between the goal sets center on the higher level thinking required for scientific research, the habits of mind that lie behind critical thinking, the hands-on opportunities provided through mentorships and internships, and focus on conceptual understanding. If these are the goals of a STEM program, then what are the instructional tools necessary to implement one? What skills must teachers exhibit to address these lofty goals?

INSTRUCTIONAL EMPHASES: TEACHING TO HIGHER LEVEL SKILLS

The higher order abilities and skills that prepare gifted students for STEM careers center on critical thinking, problem finding and problem solving, and creative responses to novel situations. These are the same lifelong learning skills on which gifted education has been centered for the last 50 years, providing the scaffolding for worthwhile learning. It is "teaching them to fish," not providing enough to eat for just a day. This constructivist approach to learning, however, requires similar approaches to be employed by the teacher, requiring long-term investment in learning new ways to think as well as to teach. Because higher order thought and creativity is not formulaic, it requires being open to the moment, asking the probing question at the right time, engaging the class in the right activity based on when students most need it, and assessing levels of functioning with regularity. For example, bridges in communities must be replaced every few generations, and the American Society of Civil Engineers sponsors an annual bridge design contest for students at the secondary level (https://bridgecontest.org), which can capture the imaginations of students, provide challenge and develop spatial understanding, and focus their attention on a specific bridge construction problem with many other impact factors to consider in their local community, such as the impact on nearby businesses, pedestrian traffic flow, and green space. Constructive teaching also requires teachers to provide students with useful models in order to have schema on which to hang their ideas. However, even useful models cannot be taught mechanically; they must be thoughtfully applied by instructors and used idiosyncratically by gifted learners, so that the greatest benefits accrue. Finally, teachers must help students understand that real thinking and problem solving is hard work, that it takes effort over time to improve, and that the outcome is frequently uncertain.

USE OF MODELS

Problem-Based Learning

Selecting models that enhance the learning of higher order process skills is also desirable because their utility has been proven in countless classrooms, and research suggests that using a few selected models over time enhances learning more strongly than eclecticism (Hillocks, 1999). Several models have proven useful to teachers in addressing the higher order skills of problem solving, and creative and critical thinking in the classroom.

PBL is a highly effective model that has been used in recent years in science classrooms to promote the skills outlined in the new standards for both science and math as well as those for developing STEM talent. First used in the medical profession to socialize doctors better to patient real-world concerns, it is now selectively employed in educational settings at elementary and secondary levels with gifted learners (Boyce, VanTassel-Baska, Burruss, Sher, & Johnson, 1997; Gallagher, 2009; Gallagher & Stepien, 1996). The technique involves several important features:

1. Students are in charge of their own learning. By working in small investigatory teams, they grapple with a real-world, unstructured problem that they have a stake in and must solve within a short period of time. Students become motivated to learn because they are in charge at every stage of the process.

2. The problem statement is ambiguous, incomplete, and yet appealing to students because of its real-world quality and the stakeholder role that they assume in it. For example, students may be given roles as scientists, engineers, politicians, or important project-based administrators whose job it is to deal with the problem expeditiously.

3. The role of the teacher is facilitative not directive, aiding students primarily through question-asking and providing additional scaffolding of the problem with new information or resources needed. The teacher becomes a metacognitive coach, urging students through probing questions to deepen their inquiry.

4. The students complete a Need to Know Board early in their investigation that allows them to plan out how they will attack the problem, first by identifying what they already know from the problem statement, what they need to know, and how they will find it out. They then can prioritize what they need to know, make assignments, and set up timelines for the next phase of work. Such an emphasis on constructed metacognitive behavior is central to the learning benefits of the approach.

These features work together then in engaging the learner in important problems that matter in their world. Many times problems are constructed around specific situations involving pollution of water or air, dangerous chemicals, spread of infectious disease, or energy source problems. Students learn that the real world is interdisciplinary in orientation, requiring the use of many different thinking skills and many different kinds of expertise in order to solve problems (Kim et al., 2012; VanTassel-Baska et al., 1998).

DESIGNING ENGINEERING CURRICULUM AND ASSESSING STUDENT PERFORMANCE

From a scientific process and experimental design view, problem- and project-based learning both create an avenue to explore scientific processes by using experimental design procedures. Specifically, both PBLs prompt students to investigate a particular topic, test a hypothesis, follow through with appropriate scientific procedures, participate in academic discussions, engage in a reanalysis of the problem, and communicate their findings to a relevant audience. Problem- and project-based learning provide a contextual setting for learning (Dailey & Cotabish, 2016). In order to work through a problem-based learning episode, students must be able to analyze, synthesize, evaluate, and create—all higher level thinking tasks, according to Anderson and Krathwohl (2000).

The following problem and its levels of complex thinking are illustrative of a problem-based learning episode: *There is a lack of mass transit into and out of a central city. You are an urban planner, given one month to come up with a viable plan. However, your resources have been used on another project, that of city beautification. A new airport is about to be built 20 miles outside of city, but negotiations are stalled. What do you do?*

Higher level skills needed to address the problem include:

1. **Analysis of what the real problem is:** Could the issue be mass transit, airport construction, beautification?
2. **Synthesis of the aspects of the problem:** Is there a creative synthesis of each facet of the problems noted?
3. **Evaluation of alternative strategies to be employed:** Can I shift funds, can I employ a transportation expert, can I deal with the airport construction?
4. **Creation of the plan of action:** The plan will need to be sold to city council members.

Another PBL scenario related to engineering is provided in Table 10.4.

As described, PBL involves presenting students with a real-world problem as described in a detailed scenario with an ill-structured problem. Students derive solutions by working through a series of process steps. The seven steps include: (1) introduction of an Ill-structured problem; (2) identification of what learners know (e.g., the "3 What's;" "Need to Know;" or "KWL Chart"); (3) gathering of information; (4) generation of possible solutions; (5) determination of the best fit solution; (6) presentation of the proposed solution; and (7) resolution/action (optional; recently proposed by Adams, Cotabish, & Dailey, 2015). Steps 2–5 are not necessarily sequential and may be conducted simultaneously as new information may redefine the original problem. Adams and colleagues (2015) proposed a seventh (optional) step, resolution/action, because an additional step may be necessary to carry out the solution. Near the conclusion of the PBL experience, the class is encouraged to reach an agreement on the best-fit solution. Ideally, the

TABLE 10.4
Engineering PBL: A Floating Bridge

Problem: Commuters traveling into the city must cross a wide river, and a structure to cross the water is needed. However, traditional bridge structures will not work in the present situation due to several characteristics: (a) the road corridor is curved, (2) the height of a bridge's support towers would be out of character with the scenic surroundings and create more noise and block views, (3) the water is deep with soft bridges, and (4) the area is known for earthquakes. You are the project engineer and must identify answers for why each characteristic is problematic to a conventional bridge structure. Furthermore, the state highway department has tasked you with exploring a design and building a prototype for a floating bridge that could be used over several similar bodies of water.

Supplies: Container filled with water (e.g., storage tub, aquarium, basin), styrofoam, glue, connectors, dowel rods, other supplies as identified.

Alignment to the NGSS for Engineering Design–Middle School: MS-ETS1-1; MS-ETS1-2; MS-ETS1-4.

For additional information: Have students research the world's longest floating bridge, which opened on Washington State Route 520. Commuters traveling west traverse the world's longest floating bridge, and then, heading back east, they cross the world's second-longest floating bridge. Washington State Department of Transportation crews opened the 7,710-foot-long bridge to westbound traffic on April 11, 2016. The venture will move all traffic, providing commuters three lanes in each direction. To avoid Lake Washington's waves, the roadway is elevated 20 feet above 77 concrete pontoons. An additional 14-foot-wide bike and pedestrian path will be on the north side.

resolution of a community-based problem will lead to action, such as a service-learning project (Dailey & Cotabish, 2016). Figure 10.1 presents the similarities and differences between the two PBLs.

Critical Thinking

In PBL, the teacher begins the instruction with essential/driving questions to direct students through the investigation. The teacher scaffolds the learning for students through labs, lectures, technology applications, and instructional activities. Throughout the investigation, students create and continuously revisit a "Need to Know" list. As the investigation concludes, students reflect and consider peer and teacher feedback to better inform their learning. As a final activity, students present their products or creations to an audience, ideally to professionals in a related field of study.

Higher level process skills require students to make nuanced judgments and interpretations about data. One effective model to teach students to enhance

Similarities

Both:
- o Provide opportunities for differentiation
- o Are open-ended in nature
- o Address 21st-century learning competencies
- o Are task driven
- o Employ entry events
- o Typically conducted in groups
- o Are student centered
- o Used as a formative assessment
- o Includes the three "What's" or "Need to Know's"
- o Involve research of subject matter
- o Spur in-depth inquiry
- o Follows steps
- o Prompt critical and creative thinking

Differences	
Problem-based learning	**Project-based learning**
Typically shorter in duration	Often longer in duration
Choice is tied to possible solutions	Frequently employs student choice throughout
Often single subject	Often interdisciplinary/integrative
Products are often in the form of solutions	Emphasis on final product
Multiple paths for solving ill-structured problem	Centered around driving questions
New-found information may redirect or pose additional questions	Final products are often presented to public audiences
Often uses case studies or fictitious scenarios to set up the problem	Typically involves real-world problem
May require an additional action step to carry out and resolve the issue(s)	Employs revision and reflection
May or may not utilize technology	Utilizes technology

Figure 10.1. ProbLem-based LearninG versus project-based LearninG. From *A Teacher's Guide to Using the Next Generation Science Standards With Gifted and Advanced Learners* (p. 99), by C. M. Adams, A. Cotabish, and D. DaiLey, 2015, New York, NY: Taylor & Francis. CopyriGht 2015 by NationaL Association for Gifted ChiLdren. Reprinted with permission.

these skills is the Ennis Model of Critical Thinking, which uses judgment and inference as the centerpiece of the critical thinking process (Ennis, 1996). Although the model has been used more extensively at the secondary level, it can be applied with gifted students at upper elementary levels with successful results. An important aspect of this model is the 12 dimensions of critical thinking Ennis (1996) derived from a study of the literature and his own philosophically trained education.

The first dimension of his model involves all aspects of interpretation, whether it is derived by inductive or deductive means. A student activity that aids the development of interpretation might be to have students study proverbs or the sayings of great writers and philosophers. Presented with a statement of import, students could be asked the following questions:

- What do the significant words mean?
- What does each line of the statement mean?
- What situations does the statement refer to?
- What ideas about life does it share?
- What new applications can you make to the idea that relate to your life and to the society as a whole today?

Another model that has proven helpful to many teachers and other educators in the application of critical thinking to real life has been the use of Richard Paul's elements of reasoning (Paul & Elder, 2008). These elements can be applied to engineering content scenarios and include the following:

1. **Purpose, Goal, or End View:** We reason to achieve some objective, to satisfy a desire, to fulfill some need. For example, if the car does not start in the morning, the purpose of my reasoning is to figure out a way to get to work. One source of problems in reasoning is traceable to "defects" at the level of purpose or goal. If our goal itself is unrealistic, contradictory to other goals we have, confused or muddled in some way, then the reasoning we use to achieve it is problematic. If we are clear on the purpose for our writing and speaking, it will help focus the message in a coherent direction. The purpose in our reasoning might be to persuade others. When we read and listen, we should be able to determine the author's or speaker's purpose.

2. **Question at Issue (or Problem to Be Solved):** When we attempt to reason something out, there is at least one question at issue or problem to be solved (if not, there is no reasoning required). If we are not clear about what the question or problem is, it is unlikely that we will find a reasonable answer, or one that will serve our purpose. As part of the reasoning process, we should be able to formulate the question to be answered or

the issue to be addressed. For example, "Why won't the car start?" or "Should libraries censor materials that contain objectionable language?"

3. **Points of View or Frame of Reference:** As we take on an issue, we are influenced by our own point of view. For example, parents of young children and librarians might have different points of view on censorship issues. The price of a shirt may seem too low to one person while it seems high to another because of a different frame of reference. Any defect in our point of view or frame of reference is a possible source of problems in our reasoning. Our point of view may be too narrow, may not be precise enough, may be unfairly biased, and so forth. By considering multiple points of view, we may sharpen or broaden our thinking. In writing and speaking, we may strengthen our arguments by acknowledging other points of view. In listening and reading, we need to identify the perspective of the speaker or author and understand how it affects the message delivered.

4. **Experiences, Data, Evidence:** When we reason, we must be able to support our point of view with reasons or evidence. Evidence is important in order to distinguish opinions from reasons or to create a reasoned judgment. Evidence and data should support the author's or speaker's point of view and can strengthen an argument. An example is data from surveys or published studies. In reading and listening, we can evaluate the strength of an argument or the validity of a statement by examining the supporting data or evidence. Experiences can also contribute to the data of our reasoning. For example, previous experiences in trying to get a car to start may contribute to the reasoning process that is necessary to solve the problem.

5. **Concepts and Ideas:** Reasoning requires the understanding and use of concepts and ideas (including definitional terms, principles, rules, or theories). When we read and listen, we can ask ourselves, "What are the key ideas presented?" When we write and speak, we can examine and organize our thoughts around the substance of concepts and ideas. Some examples of concepts are freedom, friendship, and responsibility.

6. **Assumptions:** We need to take some things for granted when we reason. We need to be aware of the assumptions we have made and the assumptions of others, and to acknowledge the importance of the beliefs that underlie people's point of view. If we make faulty assumptions, this can lead to defects in reasoning. As a writer or speaker, we make assumptions about our audience and our message. For example, we might assume that others will share our point of view, or we might assume that the audience is familiar with the First Amendment when we refer to "First

Amendment rights." As a reader or listener, we should be able to identify the assumptions of the writer or speaker.

7. **Inferences:** Reasoning proceeds by steps called inferences. An inference is a small step of the mind, in which a person concludes that something is so because of something else being so or seeming to be so. The tentative conclusions (inferences) we make depend on what we assume as we attempt to make sense of what is going on around us. For example, we see dark clouds and infer that it is going to rain; or we know the movie starts at 7:00, it is now 6:45, and it takes 30 minutes to get to the theater, so we cannot get there on time. Many of our inferences are justified and reasonable, but many are not. We need to distinguish between the raw data of our experiences and our interpretations of those experiences (inferences). Also, the inferences we make are influenced by our point of view and assumptions.

8. **Implications and Consequences:** When we reason in a certain direction, we need to look at the consequences of that direction. When we argue and support a certain point of view, solid reasoning requires that we consider the implications of following that path; what are the consequences of taking the course that we support? When we read or listen to an argument, we need to ask ourselves what follows from that way of thinking. We can also consider consequences of actions that characters in stories take. For example, if I don't do my homework, I will have to stay after school to do it; if I water the lawn, it will not wither in the summer heat.

By applying these elements systematically to different situations and events, students come to reason out both personal and real-world problems that they encounter. By converting topics to issues, students also learn the value of questioning all sides of an issue. For example, instead of having students study animal habitats from a topical perspective, why not have them debate the issue of "Should animals have rights?" or "Should we protect endangered species?"

Such a transformation of the focus for debate, discussion, and project work takes the activity to a higher level of thought and reflection. Moreover, it sets up the possibility for a dialectic that pushes the thinking of the group to a higher level as well.

ENGINEERING PROCESS METHODS

As discussed earlier, PBL provides learners with practice scenarios to process through solutions in a revisionary fashion. This process mirrors the professional processes used by engineers in the field. In the real world of engineers, there are approaches they employ to carry out their work. PBL readies students for more sophisticated processes and these processes can also be modeled in the classroom for students in grades K–12. Teachers can describe and introduce students to these methods from the field. Several engineering methods include the following:

- **The Design Method** is an instructional approach for students to work through a problem-solving process. Top down design is used in systems engineering and computer science to clearly define a problem and determine what a good solution needs to do. Many businesses fail because they are not focused on solving the right problem. A possible classroom activity practicing this process could be: "We need to build a new car" in which the teacher asks questions to determine the number of seats, how big the trunk should be, speed of the car, gas mileage, etc., to build a clearly defined list the actual construction process involves collecting individual pieces of the car either procured or designed and assembled in a bottom up fashion.
- **Design Modification** focuses on using existing components as much as possible but in a new way and with a new element. A possible activity prompt may be: *What could we add to a shopping cart to make it better?*
- **Fault Analysis** focuses on identifying everything that went wrong when a negative event occurs. A teacher may ask: Why did the power go out? Why did the plane crash? What caused the bridge to fall? Usually there were multiple contributing factors, and each should be addressed. Learning about manufacturing processes, such as the Ishikawa Method, can assist Fault Analysis by grouping failure causes by category.

Digital computers have a prominent and growing role in modern society and are involved in scientific discoveries and engineering developments, manufacturing and business transactions, communications, transportation, medical treatment, and entertainment. Computer-aided design (CAD) allows us to build and refine design, using computers. Advantages of this process include that it can be changed before construction begins. Multiple stakeholders can be designing the item at the same time (e.g., design three components of structure, electrical, plumbing in a building). CAD is used in skyscrapers and kitchen remodeling. There are several low-cost versions that students can use to design buildings or bridges in a general sense, prior to construction grade drawings. Teachers can download software to simulate drawing plans. These applications again use the skills and processes of critical thinking and PBL.

The Iterative Method helps students focus on building better prototypes. Iterative design is used to test a new concept to see if the concept works for consumers and then quickly refine the design based on testing and feedback. It can be key to PBL scenarios, as students refine solutions developed using CAD or other modeling and simulation prototypes.

Because STEM careers demand more attention to the combination of skills and abilities, many of them found in the subjects of mathematics and science, it is important to apply multiple models for instructional fidelity of curriculum implementation. Mathematical and scientific thinking form the basis for critical thinking while PBL provides the stimulus context for tasks and activities. Table 10.5 reveals the needed emphasis in these areas, as students design research projects that demonstrate their ability to think and problem solve.

CONCLUSION

Integrating PBL into engineering curriculum is an applied method that can increase engineering interest and skills among high-ability learners. As teachers develop PBL scenarios for students to explore, each problem should include related engineering content; alignment to the NGSS engineering standards; an ambiguous, open-ended problem to solve; and encourage elements of creativity to generate many original and flexible ideas with elaboration of details for solving the problem. Students grow through sophisticated learning experiences that provide opportunities to use advanced thinking processes and create advanced products. Sequencing problems to build in complexity further develop depth of understanding among high-ability learners.

In revisiting Table 10.1, we can consider the sample list of suitable curriculum problems facing thinkers in modern society. Teachers can use these ideas to develop local scenarios for engaging students and applying the instructional models of problem-based thinking, design and process methods, critical thinking skills, and best practices in writing curriculum to challenge high-ability learners in understanding engineering and developing problem-solving abilities.

Designing an engineering curriculum for gifted learners through applications of problem-based learning and other models that address high-level engineering skills appears to be the roadmap needed to proceed in enhancing STEM talent in the workplace. Teachers must be ready to undertake the challenge of engaging young learners in problem-based scenarios, materials that fascinate them, and

DESIGNING ENGINEERING CURRICULUM AND ASSESSING STUDENT PERFORMANCE

TABLE 10.5
Assessment of Research Projects of Gifted Students in STEM

	Needs Improvement	Satisfactory	Excellent
1. **Issue or problem** is clearly defined.	1	2	3
2. **Sources** are diverse.	1	2	3
3. Sources are **summarized**.	1	2	3
4. **Interview or survey questions** are included.	1	2	3
5. Interviews and/or surveys are **summarized**.	1	2	3
6. **Results** are reported appropriately.	1	2	3
7. **Interpretation of data** was appropriate.	1	2	3
8. Given the data, **reasonable conclusions** were drawn.	1	2	3
9. **Appropriate implications** were made from the conclusions.	1	2	3
10. The project paper was **mechanically competent**.	1	2	3
Strengths:			
Areas for Improvement:			

Note. This rubric is one that may be used in any STEM subject or combined class as it focuses on the thinking and problem-solving skills required for students to engage in problem-based projects. It represents the basis for project development as well as assessment. Too often student projects have not been designed how those projects will be assessed and focus on a random set of skills and behaviors. By focusing on a narrow set of thinking skills, students may demonstrate improvement as they move through a STEM curriculum K–12.

asking questions that direct their curiosity to solving problems of interest. The preschool teacher in the scenario that began the chapter said, "Give them junk and they will make products that work." This advice is also important for older gifted students, whose skills and abilities need to be challenged in the real world of materials and construction tasks.

REFERENCES

Adams, A., Cotabish, A., & Dailey, D. (2015). *A teacher's guide to using the Next Generation Science Standards with gifted and advanced learners.* Waco, TX: Prufrock Press.

Anderson, L., & Krathwohl, D. (Eds.). (2000). *A taxonomy for learning, teaching, and assessing: A revision of Bloom's taxonomy of educational objectives* (Abridged ed.). New York, NY: Longman/Pearson.

Baldi, S., Jin, Y., Green, P. J., & Herget, D. (2007). *Highlights from PISA 2006: Performance of U.S. 15-year-old students in science and mathematics literacy in an international context.* Washington, DC: U.S. Department of Education, National Center for Education Statistics.

Borman, G. D., & Hewes, G. M. (2002). The long-term effects and cost-effectiveness of success for all. *Success for All, 24,* 243–266.

Boyce, L., VanTassel-Baska, J., Burruss, J., Sher, B., & Johnson, D. (1997). A problem-based curriculum: Parallel learning opportunities for students and teachers. *Journal for the Education of the Gifted, 20,* 363–379.

California 4-H Youth Development Program. (2016). 4-H robotics: Engineering for today and tomorrow. *Junk Drawer Robotics.* Retrieved from http://4h.ucanr.edu/Projects/STEM/SET_Projects/Tech/JDR

Dailey, D., & Cotabish, A. (2016). E is for engineering education: Cultivating applied science understandings and problem-solving abilities. In B. MacFarlane (Ed.), *STEM education for high-ability learners: Designing and implementing programming* (pp. 71–84). Waco, TX: Prufrock Press.

Ennis, R. H. (1996). Critical thinking. *Argumentation, 14*(1), 48–51.

Feng, A., VanTassel-Baska, J., Quek, C., Bai, W., & O'Neill, B. (2005). A longitudinal assessment of gifted students' learning using the integrated curriculum model (ICM): Impacts and perceptions of the William and Mary language arts and science curriculum. *Roeper Review, 27,* 78–83.

Gallagher, S. A. (1989). Predictors of SAT mathematics scores of gifted male and female adolescents. *Psychology of Women Quarterly, 13,* 191–203.

Gallagher, S. A. (2009). What do you need to know? Becoming an effective PBL teacher. In B. MacFarlane & T. Stambaugh (Eds.), *Leading change in gifted education: The festschrift of Dr. Joyce VanTassel-Baska* (pp. 337–350). Waco, TX: Prufrock Press.

Gallagher, S. A., & Stepien, W. J. (1996). Content acquisition in problem-based learning: Depth versus breadth in American studies (Abstract). *Journal for the Education of the Gifted, 19,* 257–275.

Gavin, M. K., Casa, T. M., Adelson, J. L., Carroll, S. R., Sheffield, L. J., & Spinelli, A. M. (2007). Project M³: Mentoring mathematical minds: Challenging curriculum for talented elementary students. *Journal of Advanced Academics, 18,* 566–585.

Gonzales, P., Williams, T., Jocelyn, L., Roey, S., Kastberg, D., & Brenwald, S. (2008). *Highlights from TIMSS 2007: Mathematics and science achievement of U.S. fourth- and eighth-grade students in an international context.* Washington, DC: U.S. Department of Education, National Center for Education Statistics.

Grigg, W., Lauko, M., & Brockway, D. (2006). *The nation's report card: Science 2005.* Washington, DC: U.S. Government Printing Office.

Hertberg-Davis, H., & Callahan, C. M. (2008). A narrow escape: Gifted students' perceptions of Advanced Placement and International Baccalaureate programs. *Gifted Child Quarterly, 52,* 199–216.

Hillocks, G. (1999). *Ways of thinking, ways of teaching.* New York, NY: Teachers College Press.

Kim, K. H., VanTassel-Baska, J., Bracken, B. A., Feng, A., Stambaugh, T., & Bland. L. (2012). Project Clarion: Three years of science instruction in title I schools among K-third grade students. *Research in Science Education, 42,* 813–829. doi:10.1007/s11165-011-9218-5

Lawrence, B., Hinterlong, D., & Sutherland, L. (2015). Engineering education for high-ability students. In F. Dixon & S. Moon (Eds.), *The handbook of secondary gifted education* (pp. 391–416). Waco, TX: Prufrock Press.

Lubinski, D., & Benbow, C. P. (2006). Study of Mathematically Precocious Youth after 35 years: Uncovering antecedents for the development of math-science expertise. *Perspectives on Psychological Science, 1,* 316–345.

MacFarlane, B. (2016). Infrastructure of comprehensive STEM programming for advanced learners. In B. MacFarlane (Ed.), *STEM education for high-ability learners: Designing and implementing programming* (pp. 139–162). Waco, TX: Prufrock Press.

Mioduser, D., & Betzer, N. (2008). The contribution of project-based-learning to high-achievers' acquisition of technological knowledge and skills. *International Journal of Technology and Design Education, 18,* 59–77.

NGSS Lead States. (2013). *Next Generation Science Standards: For states, by states.* Washington, DC: The National Academies Press.

National Research Council. (2012). *A framework for K–12 science education: Practices, crosscutting concepts, and core ideas.* Washington, DC: The National Academies Press.

National Science Board. (2010). *Science and engineering indicators 2010.* Arlington, VA: National Science Foundation, National Center for Science and Engineering Statistics.

OECD. (2007). *PISA 2006: Science competencies for tomorrow's world* (Vol. 1). Retrieved from http://www.oecd.org/edu/school/programmeforinternationalstudentassessmentpisa/pisa2006results.htm

Paul, R., & Elder, L. (2008). *The thinker's guide for conscientious citizens on how to detect media bias and propaganda in national and world news* (4th ed.). Foundation for Critical Thinking. Retrieved from https://www.criticalthinking.org/store/products/how-to-detect-media-bias-amp-propaganda-4th-edition/167

Positive Development of Youth. (2013). *The positive development of youth: Comprehensive findings from the 4-H study of positive youth development.* Chevy Chase, MD: National 4-H Council.

Ramey, C. T., & Ramey, S. L. (1998). Early intervention and early experience. *American Psychologist, 53,* 109–120.

Rayneri, L. J., Geber, B. L., & Wiley, L. P. (2006). The relationship between classroom environment and the learning style preferences of gifted middle school students and the impact on levels of performance. *Gifted Child Quarterly, 50,* 104–118.

Robinson, A., Dailey, D., Hughes, G., & Cotabish, A. (2014). The effects of a science-focused STEM intervention on gifted elementary students' science knowledge and skills. *Journal of Advanced Academics, 25,* 189–213.

Swanson, J. D. (2006). Breaking through assumptions about low-income, minority gifted students. *Gifted Child Quarterly, 50,* 11–25.

Texas A&M AgriLife Extension Service. (2015). *The Texas 4-H roundup robotics challenge invitational.* The Texas A&M University System. Retrieved from http://texas4-h.tamu.edu/wp-content/uploads/2015/09/robotics_challenge_invitational1.pdf

VanTassel-Baska, J., Avery, L. D., Hughes, C. E., & Little, C. A. (2000). An evaluation of the implementation of curriculum innovation: The impact of William and Mary units on schools. *Journal for the Education of the Gifted, 23,* 244–272.

VanTassel-Baska, J., Bass, G., Ries, R., Poland, D., & Avery, L. D. (1998). A national study of science curriculum effectiveness with high ability students. *Gifted Child Quarterly, 42,* 200–211.

VanTassel-Baska, J., Feng, A. X., & Brown, E. (2008). A study of differentiated instructional change over three years. *Gifted Child Quarterly, 52,* 297–312.

VanTassel-Baska, J., Zuo, L., Avery, L. D., & Little, C. A. (2002). A curriculum study of gifted student learning in the language arts. *Gifted Child Quarterly, 46,* 30–44.

West, M. (2012, September). *STEM education and the workplace.* Retrieved from: http://www.chiefscientist.gov.au/wp-content/uploads/OPS4-STEMEducation AndTheWorkplace-web.pdf

DESIGNING ENGINEERING CURRICULUM AND ASSESSING STUDENT PERFORMANCE

CHAPTER 11

Assessments for K-8 Engineering

What Is Available and Advisable for Talented Students?

Ann Robinson, Kristy Kidd, and Jill L. Adelson

With the emphasis placed on engineering by the Next Generation Science Standards (NGSS), teachers and administrators are challenged to locate high-quality, rigorous curriculum to meet the needs of talented elementary students. The search for appropriate and technically sound engineering assessments is no less daunting. Elementary grade assessments are scarce but do exist. As students move through middle school and into secondary schools, more assessments are available. In fact, beginning in 2014, the National Assessment of Educational Progress (NAEP) administered the first nationwide assessment in technology and engineering literacy to students in grade 8 (National Center for Education Statistics, 2016). Spotting engineering talents or assessing engineering outcomes

DOI: 10.4324/9781003234951-15

for children in grades K–8 requires knowing where to look, what assessments can be used for specific purposes, and on what basis to select the assessments for those purposes.

In terms of the purposes of assessment, engineering tasks or measures might be used for talent spotting, to document learning outcomes, and to measure student engagement in or identity with engineering. In terms of engineering education assessments for teachers, knowledge of engineering and feelings of self-efficacy about teaching engineering affect student outcomes and should, therefore, be of interest to gifted educators implementing engineering curricula or programs (Hsu, Purzer, & Cardella, 2011).

Commonalities exist between student outcomes important in gifted education and in engineering education. For example, the match between the engineering design process and the creative problem-solving process has been explored with respect to the outcome of creativity (Cropley & Cropley, 2005; Mann, 2014) and for relevance to talented learners exposed to engineering curricula (Robinson, Kidd, & SanAntonio-Tunis, 2015). These curriculum-embedded processes require creative approaches to assessments that are technically adequate and also capture knowledge of engineering (Cunningham, Lachapelle, & Lindgren-Streigher, 2005; Duncan-Wiles, 2012), design processes (Hsu, Cardella, & Purzer, 2014), and the engagement of children in engineering (Cash, Adelson, & Robinson, 2016).

The purpose of this chapter is to:
1. describe an approach to assessing engineering-related domains, such as conceptualizations of engineers and engineering, knowledge of the engineering design process, and affective outcomes related to engineering;
2. summarize selected engineering assessments for students in grades K–8;
3. provide an example task and suggest how it meets the differentiation criteria of complexity; and
4. provide an adapted rubric for an engineering curriculum and suggest how it meets the differentiation criteria of creativity.

WHAT CAN BE ASSESSED IN K-8 ENGINEERING?

A number of engineering curricula for precollege learners, including K–8 students, exist (Brophy, Klein, Portsmore, & Rogers, 2008; Lachapelle & Cunningham, 2014). Although curriculum developers have adopted various models and the level of integration among the STEM disciplines to which the developers subscribe varies (Roehrig, Moore, Wang, & Park, 2012), the area of design is common to engineering curricula and, therefore, should be assessed

(Dym, 2004; Lewis, 2005). In addition to engineering design, other outcomes have attracted attention, such as children's conceptions of engineers and what they do and a family of affective outcomes that include attitudes, engagement, and identity with engineering. In summary, for a gifted educator seeking guidance, K–8 engineering assessments for students focus on:

- children's conceptions of engineers,
- understanding of engineering design, and
- affective measures of attitudes and engagement.

HOW CAN WE ASSESS WHAT CHILDREN THINK ABOUT ENGINEERS AND THEIR WORK?

To date, the most frequently reported assessment measure is the Draw An Engineer Test (DAET) developed by Knight and Cunningham (2004) as an adaptation of the Draw-A-Scientist Test (DAST) pioneered 40 years ago by Chambers (1983). The DAET is an open-ended task asking children to draw an engineer doing work. It has been used with young children in grades 1–5 (Capobianco, Diefes-Dux, Mena, & Weller, 2011), with a small sample of gifted children in grades 3–4 attending a summer enrichment program (Oware, Capobianco, & Diefes-Dux, 2007), and with sixth-grade African-American students. A checklist and a scoring guide have been developed to accompany the task, piloted with middle school students, and published (Fralick, Kearn, Thompson, & Lyons, 2009; Thompson & Lyons, 2008).

The open-ended task with accompanying prompts can elicit complex conceptualizations of engineers and their work. As a formative assessment tool, students are likely to find the DAET engaging to complete, and teachers can observe rich performances with the DAET and the DAET Scoring Guide. Nevertheless, as Cunningham, LaChapelle, and Lindgren-Streicher (2005) note, there are limitations to the open-ended DAET. The DAET relies on a single student-generated image of an engineer, is not easily scored, and therefore, would be difficult to implement in assessing student growth or in evaluating program efficacy.

To address those barriers to implementation, the "What is Engineering?" measure was developed. As the assessment continues to be refined and validated, "What is Engineering?" appears in more than one version, is embedded in the Engineering is Elementary curriculum units developed by the Museum of Science, Boston, and can be found on the Engineering is Elementary website (http://www.eie.org/eie-curriculum/research-instruments). In addition to the 19-item curriculum-embedded version, the researchers have developed a 37-item version that is currently undergoing an extensive validation study funded by the

National Science Foundation. All three of these versions of the assessment focused on children's conceptions of engineers and on the nature of engineering work are resources for diagnosing and documenting student learning.

HOW CAN WE ASSESS STUDENTS' UNDERSTANDING OF THE ENGINEERING DESIGN PROCESS?

The engineering design process plays a major role in grades K–8 engineering curricula. As an instructional feature, the process is iterative rather than a linear set of steps. The possibilities for assessing the engineering design process are rich. An observation protocol in the hands of a teacher prepared to teach the design process would be a welcome addition to the assessment toolbox. At present, however, the assessments of the engineering design process for children in grades K–8 rely on an analytical task that asks students to critique a pictorial representation of a child engaged in designing a container for an egg drop contest (Hsu et al., 2014), and a variety of rubrics that can be applied to children's written explanations of how they might solve a specific design problem. In the egg container task, students are shown the diagram of a child named Chris engaged in the design process and asked questions: *What was good about the process Chris used? What could be changed? What should Chris do differently next time?* The responses are scored on a 7-point scale linked to the design process used in the Engineering is Elementary model—ask, imagine, plan, create, and improve—with the addition of two more concepts—test and document. The task remains under development, but rubrics that allow teachers to score either written explanations from children or to observe them discussing the egg container design problem could move this task from a research instrument to one useful for formative assessment in classrooms. The egg container task is likely to elicit complex thinking. Thus, an accompanying rubric constructed to capture analytical thinking could meet the criterion of complexity for a differentiated assessment.

Other design tasks are available to teachers at no cost from websites, such as http://pbskids.org/designsquad. For example, we have adapted one of the Public Broadcasting Service (PBS) engineering design tasks, *Harmless Holders*. Our differentiated task appears as Figure 11.1 and is paired with an adapted rubric.

The rubric, Figure 11.2, focuses on the engineering design process and has been further differentiated for advanced learners by including the additional dimension of creativity. Although engineers collaborate on designs, this rubric is applied to an individual student's written or illustrated design process.

Can You Design a Better Can Carrier?	
Objective	Use the engineering design process as well as science content knowledge to solve a problem.
Next Generation Science Standards and Performance Expectations	Grade 2-Structure and Properties of Matter. Grade 2-Performance Expectations o 2-PS1-1-Plan and conduct an investigation to describe and classify different kinds of materials by their observable properties o 2-PS1-2-Analyze data obtained from testing different materials to determine which materials have the properties that are best suited for its intended purpose.
Materials	Paint stirrers, roll of wax paper, cardboard of various sizes, rolls of duct tape, rolls of masking tape, rubber bands, binder clips, copy paper, cardstock, six cans of soda or four bottles of soda, an empty plastic six-pack ring
Procedures	1. Introduce the Problem: Super Soft Drink has contacted us, and they have a problem. The company is getting a lot of negative publicity regarding the plastic rings used to hold their bottles and cans. It seems that research has proven that plastic rings have dramatically increased the number of sea bird deaths in the past 5 years. Super Soft Drink has to act fast. How can we help solve their problem? 2. Determine the problem as a class-Super Soft Drink needs a new can carrier for the cans and bottles. 3. Brainstorm materials that could be used for this purpose. 4. List the materials that will be available to students and ask students to work in groups to list the properties of each material. 5. Decide, as a class, which properties would be best suited for the can carrier and which available materials have those properties. 6. Share the design criteria with the class. The can carrier must support six cans OR four bottles, be easy to carry, safe for animals, and convenient to use. 7. Allow students time to plan and draw a can carrier design independently before deciding on a group design. 8. Create can carriers in groups once plans have been agreed on and a supply list has been generated for the teacher. 9. Evaluate the can carrier by testing its strength, testing the ease of carrying, discussing its animal friendly and

Figure 11.1 *Can You Design a Better Can Carrier?*

Procedures, continued	nonanimal friendly components, and discussing its convenience of use. 10. Discuss possible improvements to the designs and allow groups to make improvements to their models. 11. Evaluate improved designs.
Conclusion	Review the engineering design process and the science content needed to complete the design challenge.
Differentiation	o Increase the number of constraints to increase difficulty of the task for a talented student. o Add another dimension to the rubric focused on creativity.

Figure 11.1 *Continued.*

HOW CAN WE ASSESS IMPORTANT AFFECTIVE OUTCOMES IN ENGINEERING?

In addition to students' conceptualizations of engineers and their understanding of the engineering design process, affective outcomes are also important to the development of engineering talents. These outcomes include attitudes toward engineering and engagement while experiencing engineering activities. For example, the Engineering Identity Development Scale (EIDS) measures children's academic, school, occupational, and engineering attitudes (Capobianco, Diefes-Dux, & Habashi, 2009).

The Engineering Engagement Scale (EES) measures students' emotional and behavioral engagement when learning engineering in the classroom (Cash et al., 2016). The scale, which consists of 15 Likert scale items, was developed to parallel the Science and Math Engagement Scale (SMES) but is still undergoing field testing. The EES has two versions: one for first and second graders that includes graphic representations of the Likert scale and one for third through fifth graders that does not. The administrator averages items on the two scales for a measure of how engaged a student feels in engineering class.

Table 11.1 summarizes selected K–8 engineering assessments reviewed in this chapter.

	Novice 1	Apprentice 2	Proficient 3	Distinguished 4
Steps of the design process—ask, imagine, plan, create, test, and improve—are identified and implemented.	Student does not successfully identify or implement any steps of the engineering design process.	Student identifies and implements some steps of the engineering design process. Some aspects of identification and implementation are missing or inaccurate.	Student independently, correctly, and completely identifies and implements all five steps of the engineering design process	Student participates at proficient level and goes significantly beyond (e.g., by demonstrating an understanding of the iterative nature of the engineering design process).
Design reflects an understanding of science, technology and/or mathematical concepts.	Student does not successfully or correctly use science, technology, and/or mathematical content knowledge to support the design.	Student uses science, technology, and/or mathematical content knowledge to support the design. Not all information is used correctly, or some teacher direction may be needed.	Student correctly and completely uses the science, technology, and/or mathematical content knowledge he or she has learned to support the design.	Student participates at proficient level and goes significantly beyond (e.g., by using the content knowledge to identify questions for further investigation).

Figure 11.2. Design challenge rubric—general. Adapted from multiple rubrics, including Museum of Science, Boston (2016b); National Aeronautics and Space Administration (2013); Academy of Aerospace and Engineering (2013); eCYBERMISSION (2016); Howard County Public School System (n.d.); and Jin, Song, Shin, & Shin (2015).

DESIGNING ENGINEERING CURRICULUM AND ASSESSING STUDENT PERFORMANCE

	Novice 1	Apprentice 2	Proficient 3	Distinguished 4
Defined criteria and constraints are recognized and addressed in the design solution.	Student does not successfully recognize and address defined criteria and constraints in the design solution.	Student recognizes and addresses some defined criteria and constraints in the design solution. Student may need some teacher direction.	Student recognizes and addresses all defined criteria and constraints in the design solution.	Student participates at proficient level and goes significantly beyond (e.g., by discussing possible additional criteria and how the design plan will accommodate those criteria).
Improvements to the design are based on testing and evaluation of the design.	No evaluation of the design is conducted and no improvements are made to the design after testing.	Minimal design improvements are made after completing an adequate evaluation and testing the design. Some teacher direction may be needed.	Design improvements are made after completing a comprehensive evaluation and testing the design. Improvements reflect thoughtful deliberation.	Student participates at proficient level and goes significantly beyond (e.g., by testing the design after each improvement, analyzing the data, and using the data to make significant improvements).
Creative thinking about materials is used to develop an aesthetically pleasing, functional product, process, or system.	Student does not successfully use the materials creatively to design an aesthetically pleasing, functional product, process, or system.	Student uses minimal materials creatively to design an aesthetically pleasing, functional product, process, or system. Student may require teacher direction.	Student uses multiple materials creatively to design an aesthetically pleasing, functional product, process, or system.	Student participates at proficient level and goes significantly beyond (e.g., by inventing new, innovative uses for the various materials in the design).

Figure 11.2. Continued.

TABLE 11.1

Engineering Assessments for Students in Grades K-8

Instrument	Grade Levels	Source	Engineering Domains	Format	Psychometric Evidence
Design Process Knowledge Task	Grades 2–4	Hsu et al. (2014)	Design knowledge	Open-ended	Can detect across-group differences and pre-post differences
Draw an Engineer Test (DAET)	Grades K-12	Knight & Cunningham (2004)	Conceptions of engineers and their work	Open-ended	Pre-intervention-post with grades 2–4
Engineering Engagement Scale (EES)	Grades 1–5	Cash et al. (2016)	Emotional and behavioral engagement	Likert	Parallel to Science and Mathematics Engagement Scale with good reliability and construct validity evidence, but this subscale is being field tested
Engineering Identity Development Scale (EIDS)	Grades K-5	Capobianco et al. (2009)	Academic, school, and occupational identity and engineering aspirations	Likert	Low reliability, moderate construct reliability evidence
Importance of Engineering	Validated With Grades 3–4	Museum of Science, Boston (2016a)	Importance of engineering	Likert	Preliminary construct validity, discriminant pre-post and from engineers, cognitive interviews

DESIGNING ENGINEERING CURRICULUM AND ASSESSING STUDENT PERFORMANCE

TABLE 11.1, *CONTINUED*

Instrument	Grade Levels	Source	Engineering Domains	Format	Psychometric Evidence
STEM instruments created from TIMMS and NAEP by Harwell et al.	Grades 4-5 and Grades 6-8	Harwell et al. (2015)	Engineering knowledge	Multiple choice	Person-level reliability somewhat low; construct validity evidence
STEM Semantics Survey: Engineering subscale	Middle School	Tyler-Wood, Knezek, & Christiansen (2010)	Perceptions of engineering discipline	Likert-scale evaluation of paired adjectives	Strong internal consistency; preliminary construct validity evidence; student scored lower than teachers with STEM interest but higher than preservice teachers
What Is an Engineer (10 items)	Grades 1-5	Museum of Science, Boston (2016a)	Conceptions of engineers and their work	Open-response and yes/no items	None available at this time
What Is an Engineer (37 items)	Validated With Grades 3-4	Museum of Science, Boston (2016a)	Conceptions of engineers and their work	Open-response and yes/no items	Preliminary construct validity evidence, discriminant pre-post and from engineers, cognitive interviews

CONCLUSION

Engineering assessments for young children and emerging adolescents are at a larval stage of development. A few instruments exist, and fewer of them have been sufficiently investigated for their psychometric properties. For teachers and administrators searching for appropriate assessments of affective student outcomes, student academic progress, or for program evaluation purposes two major repositories are available (Cardella, Salsman, Purzer, & Stroebel, 2014). These include the web-based databases:

- INSPIRE (http://www.inspire-purdue.org) described in this volume in Appendix B, and
- Assessing Women and Men in Engineering (http://www.engr.psu.edu/awe).

In addition, instruments on both students and teachers can be found in the research literature (Harwell et al., 2015; Hong, Purzer, & Cardella, 2011; Yasar, Baker, Kurpius, Krause, & Roberts, 2006; Yoon, Evans, & Strobel, 2014). Assessments for the screening or identification of engineering talent in students in grades K–8 are currently little known or investigated (Robinson, Adelson, & Kidd, 2016).

As the importance of engineering increases for students in grades K–8, the strengths and limitations of existing assessments will become more evident. There is a pressing need to develop instruments with sufficient ceilings for talented young children that capture complex and broadened conceptualizations of engineers and their work, sophisticated understanding and demonstration of the engineering design process, and student engagement in the "E" in STEM.

REFERENCES

Academy of Aerospace and Engineering. (2013). *Engineering design process rubric.* Retrieved from https://aerospaceandengineeringacademy.files.wordpress.com/2015/09/engineering-design-process-rubric1.pdf

Brophy, S., Klein, S., Portsmore, M., & Rogers, C. (2008). Advancing engineering education in P–12 classrooms. *Journal of Engineering Education, 97,* 369–387.

Capobianco, B., Diefes-Dux, H. A., & Habashi, M. (2009). *Generating measures for engineering identity development.* Paper presented at the 39th Institute of

Electrical and Electronics Engineers Frontiers in Education Conference, San Antonio, TX.

Capobianco, B., Diefes-Dux, H. A., Mena, I., & Weller, J. (2011). What is an engineer? Implications of elementary school student conceptions for engineering education. *Journal of Engineering Education, 100,* 304–328.

Cardella, M. E., Salzman, N., Purzer, Ş., & Strobel, J. (2014). Assessing engineering knowledge, attitudes, and behaviors for research and program evaluation purposes. In Ş. Purzer, J. Stroebel, & M. E. Cardella (Eds.), *Engineering in pre-college settings: Synthesizing research, policy, and practices* (pp. 331–342). West Lafayette, IN: Purdue University Press.

Cash, K. M., Adelson, J. L., & Robinson, A. (2016, April). *Development of the Science and Math Engagement Scale.* Paper presented at the 2016 American Educational Research Association Annual Meeting and Exhibition, Washington, DC.

Chambers, D. W. (1983). Stereotypic images of the scientist: The Draw-A-Scientist Test. *Science Education, 67,* 255–265.

Cropley, D. J., & Cropley, A. J. (2005). Engineering creativity: A systems concept of functional creativity. In J. C. Kaufman & J. Baer (Eds.), *Creativity across the domains: Faces of the muse* (pp. 169–185). Mahwah, NJ: Lawrence Erlbaum Associates.

Cunningham, C. M., Lachapelle, C., & Lindgren-Streicher, A. (2005, June). *Assessing elementary school students' conceptions of engineering and technology.* Paper presented at the American Society of Engineering Education Annual Conference and Exposition, Portland, OR.

Duncan-Wiles, D. S. (2012). Students' awareness and perceptions of learning engineering: Content and construct validation of an instrument (Unpublished doctoral dissertation). Purdue University: West Lafayette, IN.

Dym, C. L. (2004). Design, systems, and engineering education. *International Journal of Engineering Education, 20,* 305–312.

eCYBERMISSION. (2016). *Engineering design process mission folder scorecard.* Retrieved from https://www.ecybermission.com/files/Rubric%20-%20 Engineering%20Design%20Process.pdf

Fralick, G., Kearn, J., Thompson, S., & Lyons, J. (2009). How middle schoolers draw engineers and scientists. *Journal of Science Education and Technology, 18,* 60–73.

Harwell, M., Guzey, S. S., Moreno, M., Moore, T. J., Phillips, A., & Roehrig, G. H. (2015). A study of STEM assessments in engineering, science and mathematics for elementary and middle school students. *School Science and Mathematics, 115,* 66–74.

Hong, T., Purzer, Ş., & Cardella, M. E. (2011). A psychometric re-evaluation of the Design, Engineering and Technology (DET) Survey. *Journal of Engineering Education, 100,* 800–818.

Howard County Public School System. (2013). *Assessment instrument— Engineering design pre-assessment.* Retrieved from http://transitiontocommon core.wikispaces.hcpss.org/secondary+SLOs

Hsu, M. C., Cardella, M. E., & Purzer, Ş. (2014). Assessing design. In Ş. Purzer, J. Stroebel, & M. E. Cardella (Eds.), *Engineering in pre-college settings: Synthesizing research, policy, and practices* (pp. 303–314). West Lafayette, IN: Purdue University Press.

Hsu, M. C., Purzer, Ş., & Cardella, M. E. (2011). Elementary teachers' views about teaching design, engineering, and technology. *Journal of Pre-College Engineering Education Research, 1*(2), 5.

Jin, S.-H., Song, K.-I., Shin, D. H., & Shin, S. (2015). A performance-based evaluation rubric for assessing and enhancing engineering design skills in introductory engineering design courses. *International Journal of Engineering Education, 31,* 1007–1020.

Knight, M., & Cunningham, C. M. (2004, June). *Draw an Engineer Test (DAET): Development of a tool to investigate students' ideas about engineers and engineering.* Paper presented at the American Society of Engineering Education Annual Conference and Exposition, Salt Lake City, UT.

Lachapelle, C. P., & Cunningham, C. M. (2014). Engineering in elementary schools. In Ş. Purzer, J. Stroebel, & M. E. Cardella (Eds.), *Engineering in pre-college settings: Synthesizing research, policy, and practices* (pp. 61–88). West Lafayette, IN: Purdue University Press.

Lewis, T. (2005). Coming to terms with engineering design as content. *Journal of Technology Education, 16*(2), 37–54.

Mann, E. L. (2014). Creativity assessment: A necessary criterion in K–12 engineering education. In Ş. Purzer, J. Stroebel, & M. E. Cardella (Eds.), *Engineering in pre-college settings: Synthesizing research, policy, and practices* (pp. 315–330). West Lafayette, IN: Purdue University Press.

Museum of Science, Boston. (2016a). Research instruments. *Engineering is elementary.* Retrieved from http://www.eie.org/eie-curriculum/research-instruments

Museum of Science, Boston. (2016b). Student assessments. *Engineering is elementary.* Retrieved from http://www.eie.org/eie-curriculum/student-assessments

National Aeronautics and Space Administration. (2013). *NASA exploration design challenge: Design evaluation rubric.* Retrieved from http://www.nasa.gov/ pdf/720691main_EDC_Rubric.pdf

National Center for Education Statistics. (2016). *Technology and engineering literacy assessment.* Retrieved from https://nces.ed.gov/nationsreportcard/tel

Oware, E., Capobianco, B., & Diefes-Dux, H. A. (2007, June). *Gifted students' perceptions of engineers? A study of students in a summer outreach program.* Paper presented at the American Society for Engineering Education Conference and Exposition, Honolulu, HI.

Robinson, A., Adelson, J. L., & Kidd, K. A. (2016). *A talent for tinkering: Developing talents in young low-income children through engineering curriculum.* Manuscript in preparation.

Robinson, A., Kidd, K. A., & San Antonio-Tunis, C. (2015, November). *Creative problem solving and engineering design: An innovative match for teachers and talented students.* Presentation to the Annual Convention of the National Association for Gifted Children, Phoenix, AZ.

Roehrig, G. H., Moore, T. J., Wang, H.-H., & Park, M. S. (2012). Is adding the E enough? Investigating the impact of K–12 engineering standards on the implementation of STEM integration. *School Science and Mathematics, 112,* 31–44.

Thompson, S., & Lyons, J. (2008). Engineers in the classroom: Their influence on African-American students' perceptions of engineering. *School Science and Mathematics, 108,* 197–211.

Tyler-Wood, T., Knezek, G., & Christiansen, R. (2010). Instruments for assessing interest in STEM content and careers. *Journal of Technology and Teacher Education, 18,* 341–363.

Yasar, S., Baker, D., Kurpius, S. R., Krause, S., & Roberts, C. (2006). Development of a survey to assess K–12 teachers' perceptions of engineers and familiarity with teaching design, engineering, and technology. *Journal of Engineering Education, 95,* 205–216.

Yoon, S. Y., Evans, M. G., & Strobel, J. (2014). Validation of the Teaching Engineering Self-Efficacy Scale (TESS) for K–12 teachers: A structural equation modeling approach. *Journal of Engineering Education, 103,* 463–485.

DESIGNING ENGINEERING CURRICULUM AND ASSESSING STUDENT PERFORMANCE

SECTION 4

Teacher Professional Development and Student Identification Considerations for Implementing Applied Engineering in K–8 Classrooms

 DOI: 10.4324/9781003234951-16

CHAPTER 12

Designing Professional Development for K–8 Teachers of Engineering

Alicia Cotabish, Umadevi Garimella,
and Gina Howes Boshears

Engineering education is unique in that its greatest challenge is the rapid pace of our ever-evolving world of information and communication technologies. In the field of engineering, tasks previously carried out by engineers can now be performed by technicians using computers, enabling engineers to pursue product and process innovation, explore more advanced technical and professional skills, and navigate a 21st-century global environment (Adams & Felder, 2008). Not only does this major shift in focus in the engineering profession demand a bold transformation in how engineering schools teach engineering, it creates a ripple effect in how we educate children to prepare for such a profession—and it should begin in the early grades. Before we can address what students should know in

 DOI: 10.4324/9781003234951-17

grades K–8, we need to understand how to prepare teachers to teach engineering processes and practices. In most settings, this requires a teacher professional development initiative focused on engineering education.

THE EVOLVING LANDSCAPE IN GRADES K-12 ENGINEERING EDUCATION

More than 20 years ago, Brandwein (1995) recommended that science talent development begin in the early grades and include "evocative instruction, stimulating idea-enactive, inquiry-oriented behavior consistently in the classroom" (p. 41) to increase science proneness in children. More recently, Keeley (2009) stressed the importance of science in the early grades to maximize the cumulative learning processes involved in developing science talent. Both Keeley (2009) and Goldston (2005) argued that science achievement and conceptual understanding are affected when students are not exposed to science instruction until the middle grades. Being akin to science, the same can be said for engineering talent development. Although historically the professional development of teachers who teach science-related classes has been largely geared toward science content and traditional problem solving, the Next Generation Science Standards (NGSS Lead States, 2013) require educators to teach engineering practices, which includes building models and theories about the natural world, and the key set of engineering practices that engineers use as they build models and systems. This implementation will require teachers to possess knowledge and skills in science and engineering, have the ability to teach sophisticated problem solving, create a classroom climate that is conducive to carrying out engineering simulations and student-directed learning, and institute an approach that allows students to discover innovative engineering solutions across broad social, cultural, ethical, and environmental contexts (Adams & Felder, 2008). In order to provide engineering educators with the tools necessary to prepare students for tomorrow's world of engineering, teachers must first understand the importance of changing their perspectives toward the practice of engineering education. This crucial step requires a supportive school culture that values teaching as scholarship and provides continual opportunities for educational development (Fink, Ambrose, & Wheeler, 2005). Although occasional workshops and incentives produce small, isolated improvements in learning about teaching, a systematic approach toward continual education effectively changes the norms, values, and behaviors within the profession (Fink et al., 2005). Furthermore, an effective engineering instructional development program must be appealing and relevant to both new and experienced educators (Felder, Brent, & Prince, 2011).

Duration or professional development is also reported to have an effect on teacher instruction and thereby, student achievement. In an analysis of teacher professional development programs in math and science prepared for the Council of Chief State School Officers (CCSSO), Blank, de las Alas, and Smith (2008) noted that most effects were reported in programs providing a minimum of 45 hours of professional development annually. These were usually in the form of summer institutes plus job-embedded activities throughout the year. Furthermore, Supovitz and Turner (2000) reported that teacher adoption of inquiry-based instructional practices were not evident until after 40 to 79 hours of science-specific professional development. In a review of literature, Gerard, Varma, Corliss, and Linn (2011) reported that teachers who participated in sustained professional development for more than a year were more likely to incorporate a curriculum that increased students' inquiry-based learning experiences.

DESIGNING TEACHER PROFESSIONAL DEVELOPMENT FOR ENGINEERING EDUCATION

Engineering practices are prominently included in the NGSS, as they were in an earlier developmental version of the standards, the *A Framework for K–12 Science Education* (National Research Council, 2012). To reflect the importance of understanding the human-built world and to recognize the value of integrating the teaching and learning of science, math, engineering, and technology, it is important to incorporate engineering practices within the domain of natural science. To do so, teachers will need knowledge of the ideas and practices in the disciplines of engineering, an understanding of instructional strategies, and the skills to implement the strategies in their classrooms. Enabling teachers to acquire this kind of learning will require profound changes to current systems of professional development and training for supporting teachers' learning. Professional development efforts are most effective when engineering-related examples and demonstrations are used throughout the training and when content is tailored to the needs and interests of the participants—keeping in mind each individual's level of experience (Felder et al., 2011). Engaging and appropriate content germane to a teacher's practice will increase the likelihood that the information will be useful and transferable to the participant's work environment.

One way to design an effective and coherent learning experience for teachers is to establish a system that includes identifying specific engineering processes and linking them to real-world professional development activities. Farmer and Klein-Gardner (2014) have identified standards for preparation and professional development for teachers of engineering that are aligned with current research in professional development and teaching and learning, and have been adopted by the American Society of Engineering Education. From this effort, they developed the Matrix for Assessing and Planning Engineering Professional Development (MAPEPD). We have included an adapted version of this tool that can be used to guide, assess, and modify a professional development program and related offerings.

Organized as a rubric, MAPEPD describes each process around four levels of emphasis—high emphasis, moderate emphasis, low emphasis, and no emphasis across the five standards they developed. These standards, indentified by Farmer and Klein-Gardner (2014), include:

- **Standard A:** Engineering Content and Practices
- **Standard B:** Pedagogical Content Knowledge for Teaching Engineering
- **Standard C:** Engineering as a Context for Teaching and Learning
- **Standard D:** Curriculum and Assessment
- **Standard E:** Alignment to Research, Standards, and Educational Practices

MAPEPD allows professional development facilitators to identify specific processes and levels of emphasis associated with each engineering standard. The matrix can guide professional development designers in their efforts to deliver competent engineering professional development. According to the authors (Reimers, Farmer, & Klein-Gardner, 2015) of MAPEPD, "these standards are intended to inform the design of future professional development offerings and, while not evaluative, may be used informally as a tool for describing the content and characteristics of professional development programs" (p. 45). The self-assessment component of the matrix will enable providers and consumers of engineering professional development to determine the extent to which a given program focuses on each of those standards (Reimers et al., 2015).

Tools such as these can be used to improve existing professional development programs or create new long- or short-term programs. These products can also assist facilitators with evaluating existing PD curriculum and pinpoint the specific processes and subprocesses that are appropriate for a group of teachers. For example, based on the level of participants' content background and experiences, the tool can offer a context for teaching standards in science, mathematics, language arts, reading, and other subjects.

THINGS TO KEEP IN MIND WHEN IMPLEMENTING A NEW PROFESSIONAL DEVELOPMENT INITIATIVE

Requiring that a singular new approach to teaching be adopted by all can be intimidating and perceived as criticism of one's current practice. Alternatively, designers of teacher professional development should suggest multiple strategies to fold into and develop the techniques they presently employ (Felder et al., 2011). In addition, subtle changes in professional development delivery can make a big difference. For example, rather than using a didactic lecture format to deliver professional development, facilitators will be more successful if they model their suggested techniques through participants' active engagement (Felder et al., 2011). When designing professional development, the professional development program facilitators should consist of both engineering education experts, who can offer strong content knowledge, and experts in pedagogy, who are skilled in communicating the content. The use of external facilitators can attract a broader audience, contribute expertise from outside of the school, and bring visibility and credibility to the program.

PREPARING GIFTED EDUCATORS FOR ENGINEERING EDUCATION: A STUDENT-OUTCOMES APPROACH

As with any professional development initiative, teacher learning experiences should align with student learning outcomes. With regard to meeting the specific needs of gifted and talented learners, professional development designers should be mindful of the specific and unique needs of gifted learners and how teacher professional development may support their learning. As a resource, the National Association for Gifted Children (NAGC) *Pre-K–Grade 12 Gifted Programming Standards* (2010) are arranged around student outcomes and can help guide the professional development effort. In addition, several educator-friendly tools have been developed by the organization to support student-learning outcomes. To assist educators in a student outcomes process, a needs assessment tool is available through NAGC for this very purpose (Cotabish, Shaunessy-Dedrick, Dailey, Keilty, & Pratt, 2015). The tool can facilitate a professional development self-study process, targeting educator-identified area(s) for development centered on student outcomes. For example, the self-study process requires a rating response to a set of four questions that directly relate to the implementation of the programming standards and their effects on students. At a more granular level, the process requires educators to consider evidence that documents progress in supporting the standards and engages educators in the contemplation of a variety of pathways that would ultimately impact student outcomes.

In addition to introducing a self-assessment option, facilitators of professional development should present a variety of instructional options that classroom teachers can use with high-ability learners. For example, the use of flexible grouping is one way to provide students opportunities to work with others with similar and different interests, strengths, and readiness. Also, providing teachers with examples of how to create and utilize a comprehensive and continuous scope and sequence to develop differentiated plans for high-ability and high-potential students could prove to be beneficial.

MOVING ENGINEERING EDUCATION FORWARD: THE SUPPORT OF ADMINISTRATION

The role of school administrators is critical in the implementation of any professional development initiative. With regard to institutionalizing engineering education, continuous professional development in engineering education must be an embedded expectation for all teachers who are required to teach it. To paraphrase Felder et al. (2011), creating this culture can be accomplished by:

- making it clear in position advertisements, interviews, and offer letters that participation in professional development is a requirement for the job;
- thoughtfully taking into account course assessments, peer evaluations, teacher evaluations, and student ratings when making personnel decisions;
- taking the time to recognize teachers' self-direction in improving their practice and reward them by nominating them for awards on a local, regional and national level to highlight their success; and
- always encouraging participation of the school administration to help keep them apprised of the needs of the teachers and bring to their attention the importance of both moral and financial support.

SUMMARY

Although professional development is often delivered through a workshop format, that is not the only delivery vehicle. In particular, professional development focused on engineering education may require securing outside resources and innovative approaches to find and deliver professional knowledge in the field. Other options to professional development might include reading groups (book club, journal reviews, etc.); observation of other teachers; mentoring or shadowing other teachers; teacher-led seminars; team teaching; peer coaching; lesson study; local teacher networks (including other schools in the district or nearby); collaborative curriculum planning; action research projects; formalized feedback from teams or master teachers; leadership or member involvement in professional associations including engineering societies; reflective practice sessions with peers; and inviting external engineering education providers to deliver seminars, symposiums, or workshops.

Regardless of the conduit used to deliver teacher professional development in engineering, the approach to the professional development effort should remain standards-focused and encompass activities with student outcomes in mind. Districts will want to ensure that teachers possess a common set of engineering knowledge and skills as well as secure opportunities for teachers to hone their skills. By providing a range of options and choices in engineering education, professional development designers and facilitators can accommodate for individual teacher learning needs as opposed to a one-size-fits-all strategy.

REFERENCES

Adams, R. S., & Felder, R. M. (2008). Reframing professional development: A Systems approach to preparing engineering educators to educate tomorrow's engineers. *Journal of Engineering Education, 97,* 239–240.

Blank, R. K., de las Alas, N., & Smith, C. (2008). *Does teacher professional development have effects on teaching and learning? Analysis of valuation findings from programs for mathematics and science teachers in 14 states.* Retrieved from http://www.ccsso.org/Documents/2008/Does_Teacher_Professional_Development_2008.pdf

Brandwein, P. F. (1995). *Science talent in the young expressed within ecologies of achievement* (RBDM9510). Storrs: University of Connecticut, National Research Center on the Gifted and Talented.

Cotabish, A., Shaunessy-Dedrick, E., Keilty, W., Dailey, D., & Pratt, D. (2015). *Using the NAGC pre-k–grade 12 gifted programming standards to self-assess your practice or program.* Washington, DC: National Association for Gifted Children.

Farmer C., & Klein-Gardner, S. (2014). *K–12 teachers of engineering—professional development matrix.* Retrieved from https://www.asee.org/documents/papers-and-publications/papers/outreach/K-12_Teachers_of_Engineering-Professional_Development_Matrix.pdf

Felder, R. M., Brent, R., & Prince, M. J. (2011). Engineering instructional development: Programs, best practices, and recommendations. *Journal of Engineering Education, 100,* 89–122.

Fink, L. D., Ambrose, S., & Wheeler, D. (2005). Becoming a professional engineering educator: A new role for a new era. *Journal of Engineering Education, 94,* 185–194.

Gerard, L. F., Varma, K., Corliss, S. B., & Linn, M. C. (2011). Professional development for technology-enhanced inquiry science. *Review of Educational Research, 81,* 408–448.

Goldston, D. (2005). Elementary science: Left behind? *Journal of Science Teacher Education, 16,* 185–187.

Keeley, P. (2009, April). Elementary science education in the K–12 System. *NSTA WebNews Digest.* Retrieved from http://www.nsta.org/publications/news/story.aspx?id=55954

National Association for Gifted Children. (2010). *NAGC Pre-K–Grade 12 Gifted Programming Standards: A blueprint for quality gifted education programs.* Washington, DC: Author.

National Research Council. (2012). *A framework for K–12 science education: Practices, crosscutting concepts, and core ideas.* Washington, DC: The National Academies Press.

NGSS Lead States. (2013). *Next Generation Science Standards: For states, by states.* Washington, DC: The National Academies Press.

Reimers, J. E., Farmer, C. L., & Klein-Gardner, S. S. (2015). An introduction to the standards for preparation and professional development for teachers of engineering. *Journal of Pre-College Engineering Education Research, 5*(1), 40–60. http://dx.doi.org/10.7771/2157-9288.1107

Supovitz, J. A., & Turner, H. M. (2000). The effects of professional development on science teaching practices and classroom culture. *Journal of Research in Science Teaching, 37,* 963–980.

CHAPTER 13

Integrating Engineering Into K–8 Classrooms

A Method of Identifying and Developing Strong Spatial Skills

Kinnari Atit, Kay E. Ramey, David H. Uttal,
and Paula M. Olszewski-Kubilius

Historically, giftedness has been seen as a generic, innate, intellectual quality of an individual that allows one to be successful in reasoning in all domains (e.g., Robinson, Zigler, & Gallagher, 2000). Contrary to this view, research suggests that giftedness is domain-specific and malleable (e.g., Bloom, 1985; Subotnik, Olszewski-Kubilius, & Worrell, 2011) and that it must be developed and sustained by way of training and interventions in domain-specific skills (e.g., Lubinski, 2010). In this chapter, we address identifying and developing talent in kindergarten to eighth grade (K–8) students for one specific domain—engineering.

DOI: 10.4324/9781003234951-18

The field of gifted education has devoted a great deal of research and discussion to issues surrounding the identification of academic talent in children (Johnsen, 2011). Issues that have garnered the most attention include the efficacy of the various methods of identification (IQ tests, achievement tests, nonverbal tests, and creativity tests; Lohman, 2005a; Lohman & Lakin, 2008); the use of assessments to identify academic talent equitably among racially and socio-economically diverse groups of students (Lohman, 2005b); and the predictive validity of various identification tools for forms of adult achievement (Subotnik et al., 2011)—all with the goal of generating best practices for educators. Today, schools still rely heavily on measures of general intelligence and overall achievement to identify giftedness, and consequently, many talented students could be overlooked (National Association for Gifted Children [NAGC] & Council of State Directors of Programs for the Gifted [CSDPG], 2015).

However, more recent research demonstrates the efficacy of viewing talent as situated within specific domains of practice. This talent development perspective comes with an emphasis on identifying specific skillsets that are related to achievement within particular domains (see Subotnik et al., 2011; Lubinski, 2010). Research about domain-relevant skills is more substantial and typical in athletics (Elferink-Gemser, Kannekens, Lyons, Tromp, & Visscher, 2010) and the performing arts (Subotnik & Jarvin, 2005) but has also been demonstrated for some academic fields, such as STEM (science, technology, engineering, and math) fields (Park, Lubinski, & Benbow, 2007). Subotnik et al. (2011) provided a more thorough discussion of domain-specific skills: Promoters of the talent development framework assert that a domain-specific perspective of giftedness will result in the identification of more students with talents, as well as a more successful effort at promoting domain expertise.

Substantial evidence exists that abilities can be enhanced (Herrnstein, Nickerson, deSanchez, & Swets, 1986; Sternberg, 1988; Uttal et al., 2013) and that high abilities within a domain are necessary but not sufficient for generating expertise or exceptional productivity. Without opportunities to learn from skilled instructors and develop psychosocial skills such as persistence, domain-specific skills may develop too slowly or even counterproductively (Subotnik & Jarvin, 2005). For example, in engineering, learning to interpret blueprints incorrectly can lead to constructing poor quality structures. In many domains, training and instruction from a young age is important for an individual to reach an exceptional level in adulthood (Subotnik et al., 2011).

The United States hopes to develop America's future scientists, technologists, engineers, and mathematicians in order to be competitive in a growing global economy (Friedman, 2005; Jolly, 2009). One way to reach this goal is to introduce all STEM disciplines (not just the math and science traditionally taught in K–8 classrooms) early on in students' education. Engineering, a distinct STEM domain, has principles and concepts that can be easily adapted for elementary and middle school students and for different ability levels. Furthermore, engaging in engineering activities develops skills necessary to succeed in the discipline and important for success in the other STEM domains (e.g., spatial skills; Shea, Lubinski, & Benbow, 2001). The adaptability of the lessons and the fostering of skills critical to future STEM success makes engineering an ideal discipline to integrate into K–8 curriculum (e.g., Museum of Science, Boston, 2015; Samuels & Seymour, 2015).

Therefore, we will first discuss some of the advantages of introducing all students, gifted and those not identified, to engineering early on. Second, we will talk about a specific set of skills that are important for engineering—spatial skills—and how they can be assessed in students. Third, we will discuss how engineering can be integrated into K–8 curriculum and how lessons can be adapted to challenge students of different ability levels. And finally, we will consider the value of adding spatial skills to the traditional measures used to identify academically talented students.

THE IMPORTANCE OF EARLY ENGINEERING LEARNING

Grades K–8 engineering learning activities have become increasingly popular in recent years, both in and outside of school (e.g., Honey & Kanter, 2013; Martinez & Stager, 2013; National Academy of Engineering [NAE] & National Research Council [NRC], 2009; NRC, 2012; National Science Foundation [NSF], 2012; Resnick & Rosenbaum, 2013). In schools, they have become a formal part of the Next Generation Science Standards (NGSS; NAE & NRC, 2009; NGSS Lead States, 2013; NRC, 2012), while out-of-school Makerspaces, tinkering studios, engineering camps, and afterschool coding or robotics clubs have grown greatly in popularity (e.g., Honey & Kanter, 2013; Martinez & Stager, 2013; Resnick & Rosenbaum, 2013).

These early engineering learning interventions are beneficial for a number of reasons. First, they spark student interest and provide opportunities to learn about engineering, potentially increasing the number of students interested in and able to pursue careers in engineering and other STEM disciplines (Mann, Mann, Strutz, Duncan, & Yoon, 2011; National Science Board [NSB], 2010). Second, they cultivate skills and habits of mind that are valuable both in engineering and in other domains, including 21st-century skills, such as collaboration, creativity, self-management, communication, and systems thinking (Hilton, 2010; Katehi,

Pearson, & Feder, 2009; Mann et al., 2011). Many within the field of gifted education find that these skills are also attributes of gifted students (Mann et al., 2011). Lastly, they provide a vehicle for teaching and improving spatial skills, a set of skills that are important in STEM (Uttal & Cohen, 2012) and in everyday reasoning and problem solving (Gauvain, 1993), but are currently undervalued and underdeveloped in K–8 schooling (e.g., NRC, 2006; Newcombe, Uttal, & Sauter, 2013).

WHAT ARE SPATIAL SKILLS?

Traditional definitions characterize spatial skills as one type of ability. For example, an early definition defines them as the ability to "search the visual field, apprehending the forms, shapes, and positions of objects as visually perceived, forming mental representations of those forms, shapes, and positions, and manipulating such representations 'mentally'" (Carroll, 1993, p. 304). Although often presented as a unitary construct, there are actually several different kinds of spatial skills (Guilford & Lacey, 1947; McGee, 1979; Newcombe & Shipley, 2015; Thurstone & Thurstone, 1941). Some common spatial skills include skills for navigation (e.g., Weisberg, Schinazi, Newcombe, Shipley, & Epstein, 2014), perspective taking (envisioning the world from a different point of view; e.g., Hegarty & Waller, 2004), and disembedding (identifying a figure within a complex array of other figures; e.g., Witkin, Oltman, Raskin, & Karp, 1971).

Not only is there more than one kind of spatial skill, but performance of one skill does not necessarily predict performance on another type. For example, Atit, Shipley, and Tikoff (2013) showed that the skills used to visualize the rotation of an object are distinct from the skills used to visualize the bending or un-bending of an object. Therefore, a student who may be good at one type of spatial skill may not be good at all spatial tasks.

Furthermore, recent research suggests that spatial skills may be domain specific (Uttal & Cohen, 2012). For example, Jee, Gentner, Forbus, Sageman, and Uttal (2009) found that geoscience students outperformed psychology students on cross-sectioning or penetrative thinking tasks involving images of rock formations, but there was no difference between the two groups on a spatially similar task involving layers of fruit or lasagna. Similar patterns have been found in other STEM disciplines, such as chemistry (Stieff, Hegarty, & Dixon, 2010), and dentistry (Hegarty, Keehner, Khooshabeh, & Montello, 2009). As spatial skills can be improved with training and practice (Uttal et al., 2013), this research indicates that perhaps these skills should be developed within the context of a specific domain.

SPATIAL SKILLS AND MEASURES RELEVANT TO ENGINEERING

As the multiplicity of spatial skills has been recognized, many researchers have focused their efforts on understanding which skills are important for which disciplines (e.g., Atit, Gagnier, & Shipley, 2015; Baartmans & Sorby, 1996, 1999; Stieff, 2013). Work in engineering has suggested that there are three major kinds of spatial skills that are important for success in the field: *mental rotation, spatial visualization,* and *two-dimensional (2-D) to three-dimensional (3-D) mental transformation* (e.g., Hegarty, 1992, 2004; Hsi, Linn, & Bell, 1997; Sorby, Casey, Veurink, & Dulaney, 2013). Mental rotation is the ability to mentally rotate a 2-D or 3-D figure (Linn & Petersen, 1985; Shepard & Metzler, 1971). Spatial visualization is the ability to piece together objects into more complex configurations or visualize and mentally transform them (Linn & Petersen, 1985; Newcombe & Shipley, 2015). For example, the ability to mentally fold and unfold an object requires spatial visualization skills (e.g., Harris, Newcombe, & Hirsh-Pasek, 2013; Milivojevic, Johnson, Hamm, & Corballis, 2003). Lastly, 2-D to 3-D transformation involves visualizing a 2-D representation in 3-D and vice versa (e.g., Atit, Weisberg, Shipley, & Newcombe, 2016; Sorby, 2009).

Before an object is ever created, an engineer must be able to visualize the object and draw its plans (Baartmans & Sorby, 1996). Completing this task requires using mental rotation, spatial visualization, and 2-D to 3-D transformation skills. In engineering, the standard drawing layout typically includes orthographic projections of top, front, right-side, or cross-section views of the object, and also an isometric, or corner view, of the object. In order to draw the different views for an object, the engineer needs to be able to visualize what each side of the object looks like. This requires mental rotation skills and spatial visualization skills (mentally unfolding the object and visualizing the individual pieces). Lastly, in order to build the object, the engineer needs to be able to interpret the 2-D drawings of the structure and visualize it in 3-D, displaying 2-D to 3-D mental transformation skills (Baartmans & Sorby, 1996).

Although many assessments have been created to measure these specific spatial thinking skills in adults (e.g., Ekstrom, French, Harman, & Dermen, 1976; Peters et al., 1995; Titus & Horsman, 2009), only a few have been created for K–8 students. Tests of mental rotation that have been found to be suitable for children include the chronometric mental rotation test created by Petra Jansen and her colleagues (Lehmann & Jansen, 2012; Kaltner & Jansen, 2014) and the Mental Rotation Subtest from the Chicago Primary Mental Abilities Test (Thurstone & Thurstone, 1941). A test of spatial visualization for children is the Mental Folding Test for Children (Harris et al., 2013), and a test of 2-D to 3-D mental transformation is the Diagrammatic Representations Test (DRT; Frick & Newcombe, 2015). As indicated by the small number of spatial tests developed for K–8 students, much more research in this area is required. Yet, these few tests can help us begin to identify students with strong engineering-relevant spatial skills.

CULTIVATING SPATIAL SKILLS THROUGH ENGINEERING LEARNING ACTIVITIES

Given the strong connection between engineering and spatial skills (e.g., Hsi et al., 1997; Sorby, 1999, 2009; Sorby & Baartmans, 2000; Sorby et al., 2013; Tseng & Yang, 2011), parents, educators, and researchers should consider how to foster the development of these skills through engineering activities at an early age (e.g., Newcombe, 2010). Research on cognitive development and engineering learning provides insight into how and when we might help students cultivate their spatial skills. For example, object manipulation, in the form of puzzle play or manual rotation, has been shown to improve preschoolers' spatial transformation or mental rotation skills (Levine, Ratliff, Huttenlocher, & Cannon, 2012; Ping, Ratliff, Hickey, & Levine, 2011). Similarly, engaging young children in talk and gesture about spatial ideas may improve spatial skills (Ping et al., 2011; Pruden, Levine, & Huttenlocher, 2011). To the extent that engineering activities provide opportunities for discussing and manipulating spatial objects and ideas, we might expect these activities to similarly improve learners' spatial skills.

However, not all engineering learning activities are the same. There are three general categories of engineering learning activities, which vary in terms of both the goals and structure of the engineering task (Ramey & Uttal, in press). The first are construction kit activities, such as LEGO, K'NEX, or Snap Circuits, in which students build devices from diagrammatic instructions (Ramey & Uttal, in press). The second category is engineering design activities. These are included in many of the school engineering curricula, such as Project Lead the Way (https://www. pltw.org) and Engineering is Elementary (http://www.eie.org). They encourage students to walk through the steps of the engineering design process (e.g., ask, imagine, plan, create, and improve) to achieve a predetermined goal, given specific material constraints (e.g., Brophy, Klein, Portsmore, & Rogers, 2008; Museum of Science, Boston, 2015). Finally, the third category of activities are tinkering or making activities. These activities are most often seen in Makerspaces, maker clubs, and museum-based tinkering studios. They are modeled after the Maker Movement and are grounded in constructionist principles of learning and the desire to integrate new tools and technologies into design (Papert, 1980; Resnick & Rosenbaum, 2013). Of the three types of engineering activities, they are the most open-ended and creative, focusing on the use of particular tools or skills instead of predetermined goals or design processes (Martinez & Stager, 2013; Resnick & Rosenbaum, 2013; Vossoughi, Escudé, Kong, & Hooper, 2013).

Different activities cultivate different spatial skills. For example, making, tinkering, and engineering design activities all encourage learners to visualize, communicate, and create novel spatial configurations (Ramey & Uttal, in press). In contrast, construction kits necessitate attention to specific spatial relations between objects and diagrammatic instructions (Ramey & Uttal, in press). Collaborative engineering learning activities allow learners to use spatial talk and gesture to think through and share spatial ideas, while activities involving design sketching or computer-aided design (CAD) modeling help students learn to use domain-specific representations to think through spatial ideas.

Different types of engineering learning activities also have different affordances for catering to gifted students. Of the different types of engineering activities, tinkering or making activities are perhaps the best suited to gifted learners, because they afford the most creativity and self-direction. These activities tend to both necessitate and cultivate the types of skills gifted students have, such as systems thinking, creativity, optimism, collaboration, communication, and attention to ethical considerations (Duke University Talent Identification Program, n.d.; Katehi et al., 2009; Mann et al., 2011).

However, through differentiated instruction, the other two types of engineering learning activities may also have their place in gifted education. For example, typical construction kits come with diagrammatic instructions drawn in 3-D perspective, because they tend to be the easiest to understand and work from. However, as previously discussed, when professional engineers or architects make or work from construction diagrams (i.e., blueprints), they tend to work from 2-D orthographic projections. Translating these types of diagrams into 3-D objects requires more complex spatial reasoning. Thus, one way to differentiate instruction for construction kits would be to present spatially talented students with more complex, orthographic diagrams to work from. Taking this a step further, instructors might consider having gifted students design their own structures using construction kit materials and then draw blueprints of their designs, so that another student could replicate them.

Finally, within engineering design activities, educators might consider either pushing gifted students' creativity by removing some of the task constraints or challenging their reasoning skills by adding additional constraints. For example, one common engineering design activity is making and adjusting blades for a wind turbine, using cardboard or balsawood, so that the turbine can lift a certain number of weights or generate a certain amount of electricity. Unfortunately, using only cardboard or balsa wood and a premade turbine base, students are limited to flat blades and a conventional turbine structure. Removing material constraints and allowing students to, for example, 3-D print curved blades or design their own turbine base would allow for greater design creativity. Similarly, increasing the number of weights the turbine has to lift or limiting the number of blades students can use creates a more challenging reasoning problem.

IDENTIFYING STUDENTS WITH STRONG SPATIAL SKILLS

Traditionally, spatial skills are not assessed when identifying gifted students. Generally, programs for academically talented students assess only verbal and quantitative abilities (e.g., Benbow & Lubinski, 1996; Benbow & Stanley, 1996). Students with a quantitative tilt (i.e., stronger quantitative skills than verbal skills) gravitate toward STEM disciplines and are more likely to pursue STEM careers as adults (e.g., Achter, Lubinski, Benbow, & Eftekhari-Sanjani, 1999; Lubinski, Webb, Morelock, & Benbow, 2001). Research has shown that including an assessment of spatial skills in addition to quantitative and verbal measures reveals more students that pursue STEM fields in the future than do quantitative skills alone (e.g., Shea et al., 2001; Wai, Lubinski, & Benbow, 2009). Thus, assessing for exceptional spatial skills in younger students could reveal earlier which students could truly benefit from engaging in more challenging STEM activities.

CONCLUSION

Given the evidence supporting the relationship between strong spatial skills and success in STEM disciplines (e.g., Wai et al., 2009; Lubinski, 2010) and the current need to develop STEM talent in the United States (e.g., Committee on Integrated STEM Education, NAE, & NRC, 2014), it is critical that spatial skill assessments be added to identification batteries for gifted and talented students.

Recognizing students with this kind of talent could unveil the next generation of STEM professionals. However, spatial talent assessment does not need to be limited to spatial tests. It can also include curricular activities, such as engineering activities, that both cultivate and help identify spatial talent. Using activities to identify talent is especially important for students from low-income or diverse cultural backgrounds, who have potential but may, for various reasons, be less likely to demonstrate it on tests.

Additionally, teachers should offer activities and curricula that require and develop spatial skills. Noticing students who could be spatially talented and adjusting lesson plans to make sure that they are challenged can help cultivate the skills and interest necessary to pursue STEM domains later on. Moreover, it could increase motivation in school overall.

Lastly, developing spatial skills should not just be a focus for students who show advanced ability or interest in the area, but they should be fostered and developed in all students. Because spatial skills are malleable (Uttal et al., 2013), those students who do not show a propensity early on can develop stronger skills with more exposure and practice. Integrating engineering concepts and activities into the curriculum provides teachers with a vehicle for engaging students of different ability levels, a method for improving this skillset, and a means to identify children with exceptional skills or interest. Therefore, we recommend introducing engineering early in K–8 curriculum, as it can benefit both spatially talented students and students with weaker spatial skills.

REFERENCES

Achter, J. A., Lubinski, D., Benbow, C. P., & Eftekhari-Sanjani, H. (1999). Assessing vocational preferences among gifted adolescents adds incremental validity to abilities: A discriminant analysis of educational outcomes over a 10-year interval. *Journal of Educational Psychology, 91,* 777.

Atit, K., Gagnier, K., & Shipley, T. F. (2015). Student gestures aid penetrative thinking. *Journal of Geoscience Education, 63,* 66–72.

Atit, K., Shipley, T. F., & Tikoff, B. (2013). Twisting space: Are rigid and non-rigid mental transformations separate spatial skills? *Cognitive Processing, 14,* 163–173.

Atit, K., Weisberg, S., Shipley, T. F., & Newcombe, N. (2016). *Reading topographic maps: Understanding 3-D information from 2-D representations.* Manuscript submitted for publication.

Baartmans, B. G., & Sorby, S. A. (1996). Making connections: Spatial skills and engineering drawings. *Mathematics Teacher, 89,* 348–357.

Benbow, C. P. & Lubinski, D. (1996). *Intellectual Talent: Psychometric and Social Issues.* Baltimore, MD: Johns Hopkins University Press.

Benbow, C. P., & Stanley, J. C. (1996). Inequity in equity: How "equity" can lead to inequity for high-potential students. *Psychology, Public Policy, and Law, 2,* 249.

Bloom, B. S. (1985). *Developing talent in young people.* New York, NY: Ballantine Books.

Brophy, S., Klein, S., Portsmore, M., & Rogers, C. (2008). Advancing engineering education in P–12 classrooms. *Journal of Engineering Education, 97,* 369.

Carroll, J. B. (1993). *Human cognitive abilities: A survey of factor-analytic studies.* New York, NY: Cambridge University Press.

Committee on Integrated STEM Education, National Academy of Engineering, & National Research Council. (2014). *STEM integration in K–12 education: Status, prospects, and an agenda for research.* Washington, DC: The National Academies Press.

Duke University Talent Identification Program. (n.d.). *Gifted characteristics.* Retrieved from https://tip.duke.edu/node/99

Ekstrom, R. B., French, J. W., Harman, H. H., & Dermen, D. (1976). *Manual for kit of factor referenced cognitive tests.* Princeton, NJ: Educational Testing Service.

Elferink-Gemser, M. T., Kannekens, R., Lyons, J., Tromp, Y., & Visscher, C. (2010). Knowing what to do and doing it: Differences in self-assessed tactical skills of regional, sub-elite, and elite youth field hockey players. *Journal of Sports Sciences, 28,* 521–528.

Frick, A., & Newcombe, N. S. (2015). Young children's perception of diagrammatic representations. *Spatial Cognition & Computation, 15,* 227–245.

Friedman, T. L. (2005). *The world is flat: A brief history of the twenty-first century.* New York, NY: Farrar, Straus and Giroux.

Gauvain, M. (1993). The development of spatial thinking in everyday activity. *Developmental Review, 13*(1), 92–121.

Guilford, J. P., & Lacey, J. L. (1947). Printed classification tests, A.A.F. In *Army Air Force Aviation Psychology Program Research Reports, No. 5* (pp. 931). Washington, DC: U.S. Government Printing Office.

Harris, J., Newcombe, N. S., & Hirsh-Pasek, K. (2013). A new twist on studying the development of dynamic spatial transformations: Mental paper folding in young children. *Mind, Brain, and Education, 7,* 49–55.

Hegarty, M. (1992). Mental animation: inferring motion from static displays of mechanical systems. *Journal of Experimental Psychology: Learning, Memory, and Cognition, 18,* 1084–1102.

TEACHER PROFESSIONAL DEVELOPMENT AND STUDENT IDENTIFICATION

Hegarty, M. (2004). Mechanical reasoning by mental simulation. *Trends in cognitive sciences, 8,* 280–285.

Hegarty, M., Keehner, M., Khooshabeh, P., & Montello, D. R. (2009). How spatial abilities enhance, and are enhanced by, dental education. *Learning and Individual Differences, 19*(1), 61–70.

Hegarty, M., & Waller, D. (2004). A dissociation between mental rotation and perspective-taking spatial abilities. *Intelligence, 32*(2), 175–191.

Herrnstein, R. J., Nickerson, R. S., de Sanchez, M., & Swets, J. A. (1986). Teaching thinking skills. *American Psychologist, 41,* 1279.

Hilton, M. (2010). *Exploring the intersection of science education and 21st century skills: A workshop summary.* Washington, DC: The National Academies Press.

Honey, M., & Kanter, D. E. (Eds.). (2013). *Design, make, play: Growing the next generation of STEM innovators.* New York, NY: Routledge.

Hsi, S., Linn, M. C., & Bell, J. E. (1997). The role of spatial reasoning in engineering and the design of spatial instruction. *Journal of Engineering Education, 86,* 151–158.

Jee, B., Gentner, D., Forbus, K., Sageman, B., & Uttal, D. H. (2009, July). Drawing on experience: Use of sketching to evaluate knowledge of spatial scientific concepts. In N. Taatgen & H. van Rijk (Eds.), *Proceedings of the 31st Annual Conference of the Cognitive Science Society* (pp. 2499–2504). Amsterdam, The Netherlands: Cognitive Science Society.

Johnsen, S. K. (2011). *Identifying gifted students: A practical guide* (2nd ed.). Waco, TX: Prufrock Press.

Jolly, J. L. (2009). The National Defense Education Act, current STEM initiative, and the gifted. *Gifted Child Today, 32,* 50–53.

Kaltner, S., & Jansen, P. (2014). Mental rotation and motor performance in children with developmental dyslexia. *Research in Developmental Disabilities, 35,* 741–754.

Katehi, L., Pearson, G., & Feder, M. (2009). *Engineering in K–12 education: Understanding the status and improving the prospects.* Washington, DC: The National Academies Press.

Lehmann, J., & Jansen, P. (2012). The influence of juggling on mental rotation performance in children with spina bifida. *Brain and Cognition, 80*(2), 223–229.

Levine, S. C., Ratliff, K. R., Huttenlocher, J., & Cannon, J. (2012). Early puzzle play: a predictor of preschoolers' spatial transformation skill. *Developmental Psychology, 48,* 530–542.

Linn, M. C., & Petersen, A. C. (1985). Emergence and characterization of sex differences in spatial ability: A meta-analysis. *Child Development, 56,* 1479–1498.

Lohman, D. F. (2005a). The role of nonverbal ability tests in identifying academically gifted students: An aptitude perspective. *Gifted Child Quarterly, 49,* 111–138.

Lohman, D. F. (2005b). *Identifying academically talented minority students.* Storrs: University of Connecticut, National Research Center on the Gifted and Talented.

Lohman, D. F., & Lakin, J. (2008). Nonverbal test scores as one component of an identification system: Integrating ability, achievement, and teacher ratings. In J. VanTassel-Baska (Ed.), *Alternative assessments with gifted and talented students* (pp. 41–66). Waco, TX: Prufrock Press.

Lubinski, D. (2010). Spatial ability and STEM: A sleeping giant for talent identification and development. *Personality and Individual Differences, 49,* 344–351.

Lubinski, D., Webb, R. M., Morelock, M. J., & Benbow, C. P. (2001). Top 1 in 10,000: A 10-year follow-up of the profoundly gifted. *Journal of Applied Psychology, 86,* 718–729.

Mann, E. L., Mann, R. L., Strutz, M. L., Duncan, D., & Yoon, S. Y. (2011, August). Integrating engineering into K–6 curriculum: Developing talent in STEM disciplines. *Journal of Advanced Academics, 22,* 639–658. doi:10.1177/1932202X11415007

Martinez, S. L., & Stager, G. (2013). *Invent to learn: Making, tinkering, and engineering in the classroom.* Torrance, CA: Constructing Modern Knowledge Press.

McGee, M. G. (1979). Human spatial abilities: Psychometric studies and environmental, genetic, hormonal, and neurological influences. *Psychological Bulletin, 86,* 889–918.

Milivojevic, B., Johnson, B. W., Hamm, J. P., & Corballis, M. C. (2003). Non-identical neural mechanisms for two types of mental transformation: event-related potentials during mental rotation and mental paper folding. *Neuropsychologia, 41*(10), 1345–1356.

Museum of Science, Boston. (2016). *Engineering is elementary.* Retrieved from http://www.eie.org

National Academy of Engineering, & National Research Council. (2009). *Engineering in K–12 education: Understanding the status and improving the prospects.* Washington, DC: The National Academies Press.

National Association for Gifted Children, & Council of State Directors of Programs for the Gifted. (2015). *State of the states in gifted education 2014–2015.* Washington, DC: NAGC.

National Research Council. (2006). *Learning to think spatially.* Washington, DC: The National Academies Press.

TEACHER PROFESSIONAL DEVELOPMENT AND STUDENT IDENTIFICATION

National Research Council. (2012). *A framework for K–12 science education: Practices, crosscutting concepts, and core ideas.* Washington, DC: The National Academies Press.

National Science Board. (2010). *Preparing the next generation of STEM innovators: Identifying and developing our nation's human capital* (Report No. NSB 10-33). Arlington, VA: National Science Foundation.

National Science Foundation. (2012). *CISE strategic plan for broadening participation.* Retrieved from http://www.nsf.gov/cise/oad/cise_bp.jsp

Newcombe, N. S. (2010). Picture this: Increasing math and science learning by improving spatial thinking. *American Educator, 34*(2), 29–43.

Newcombe, N. S., & Shipley, T. F. (2015). Thinking about spatial thinking: New typology, new assessments. In J. S. Gero (Ed.), *Studying visual and spatial reasoning for design creativity* (pp. 179–192). Netherlands: Springer. doi:10.1007/978-94-017-9297-4_10

Newcombe, N. S., Uttal, D. H., & Sauter, M. (2013). Spatial development. In P. Zelazo (Ed.), *The Oxford handbook of developmental psychology* (Vol. 1, pp. 564–590). New York, NY: Oxford University Press.

NGSS Lead States. (2013). *Next Generation Science Standards: For states, by states.* Washington, DC: The National Academies Press.

Papert, S. (1980). *Mindstorms: Children, computers, and powerful ideas.* New York, NY: Basic Books.

Park, G., Lubinski, D., & Benbow, C. P. (2007). Contrasting intellectual patterns predict creativity in the arts and sciences: tracking intellectually precocious youth over 25 years. *Psychological Science, 18,* 948–952. doi:10.1111/j.1467-9280.2007.02007.x

Peters, M., Laeng, B., Latham, K., Jackson, M., Zaiyouna, R., & Richardson, C. (1995). A redrawn Vandenberg and Kuse mental rotations test-different versions and factors that affect performance. *Brain and Cognition, 28*(1), 39–58.

Ping, R., Ratliff, K., Hickey, E., & Levine, S. C. (2011, July). Using manual rotation and gesture to improve mental rotation in preschoolers. In L. Carlson, C. Hoelscher, & T. F. Shipley, *Proceedings of the 33rd Annual Meeting of the Cognitive Science Society* (pp. 459–464). Austin, TX: Cognitive Science Society.

Pruden, S. M., Levine, S. C., & Huttenlocher, J. (2011). Children's spatial thinking: Does talk about the spatial world matter? *Developmental Science, 14,* 1417–1430.

Ramey, K. E., & Uttal, D. H. (in press). *Making sense of space: Distributed spatial sensemaking in a middle school summer engineering camp.* Manuscript submitted for publication.

Resnick, M., & Rosenbaum, E. (2013). Designing for tinkerability. In M. Honey, & D. Kanter (Eds.), *Design, make, play: Growing the next generation of STEM innovators* (pp.163–181). New York, NY: Routledge.

Robinson, N. M., Zigler, E., & Gallagher, J. J. (2000). Two tails of the normal curve: Similarities and differences in the study of mental retardation and giftedness. *American Psychologist, 55,* 1413.

Samuels, K., & Seymour, R. (2015). The middle school curriculum: Engineering anyone? *Technology and Engineering Teacher, 74*(6), 8–12.

Shea, D. L., Lubinski, D., & Benbow, C. P. (2001). Importance of assessing spatial ability in intellectually talented young adolescents: A 20-year longitudinal study. *Journal of Educational Psychology, 93,* 604.

Shepard, R. N., & Metzler, J. (1971). Mental rotation of three-dimensional objects. *Science, 171,* 701–703. Retrieved from http://www.cs.virginia.edu/~weimer/1120/ps/ps3/mental-rotation.pdf

Sorby, S. (1999). Developing 3-D spatial visualization skills. *Engineering Design Graphics Journal, 63*(2), 21–32.

Sorby, S. A. (2009). Educational research in developing 3-D spatial skills for engineering students. *International Journal of Science Education, 31,* 459–480.

Sorby, S. A., & Baartmans, B. J. (2000). The development and assessment of a course for enhancing the 3-D spatial visualization skills of first year engineering students. *Journal of Engineering Education, 89,* 301–307.

Sorby, S., Casey, B., Veurink, N., & Dulaney, A. (2013). The role of spatial training in improving spatial and calculus performance in engineering students. *Learning and Individual Differences, 26,* 20–29.

Sternberg, R. J. (1988). *The nature of creativity: Contemporary psychological perspectives.* New York, NY: Cambridge University Press.

Stieff, M. (2013). Sex differences in the mental rotation of chemistry representations. *Journal of Chemical Education, 90,* 165–170.

Stieff, M., Hegarty, M., & Dixon, B. (2010). Alternative strategies for spatial reasoning with diagrams. *Diagrams, 6170,* 115–127.

Subotnik, R. F., & Jarvin, L. (2005). Beyond expertise: Conceptions of giftedness as great performance. In R. J. Sternberg & J. E. Davidson (Eds.), *Conceptions of giftedness* (2nd ed., pp. 343–357). New York, NY: Cambridge University Press.

Subotnik, R. F., Olszewski-Kubilius, P., & Worrell, F. C. (2011). Rethinking giftedness and gifted education a proposed direction forward based on psychological science. *Psychological Science in the Public Interest, 12,* 3–54.

Thurstone, L. L., & Thurstone, T. G. (1941). Factorial studies of intelligence. *Psychometric Monographs, 2,* 1–94.

Titus, S., & Horsman, E. (2009). Characterizing and improving spatial visualization skills. *Journal of Geoscience Education, 57,* 242–254.

Tseng, T., & Yang, M. (2011). *The role of spatial-visual skills in a project-based engineering design course.* Paper presented at the 118th American Society for Engineering Education Annual Conference and Exposition, Vancouver, BC, Canada.

Uttal, D. H., & Cohen, C. A. (2012). Spatial thinking and STEM education: When, why and how? *Psychology of Learning and Motivation, 57,* 147–181.

Uttal, D. H., Meadow, N. G., Tipton, E., Hand, L. L., Alden, A. R., Warren, C., & Newcombe, N. S. (2013). The malleability of spatial skills: A meta-analysis of training studies. *Psychological Bulletin, 139,* 352–402. doi:10.1037/a0028446

Vossoughi, S., Escudé, M., Kong, F., & Hooper, P. (2013). *Tinkering, learning & equity in the after-school setting.* Paper presented at the FabLearn III Digital Fabrication in Education Conference, Stanford, CA. Retrieved from http://fablearn.stanford.edu/2013/wp-content/uploads/Tinkering-Learning-Equity-in-the-After-school-Setting.pdf

Wai, J., Lubinski, D., & Benbow, C. (2009). Spatial ability for STEM domains: Aligning over 50 years of cumulative psychological knowledge solidifies its importance. *Journal of Educational Psychology, 101,* 817–835. doi:10.1037/a0016127

Weisberg, S. M., Schinazi, V. R., Newcombe, N. S., Shipley, T. F., & Epstein, R. A. (2014). Variations in cognitive maps: Understanding individual differences in navigation. *Journal of Experimental Psychology: Learning, Memory, and Cognition, 40,* 669.

Witkin, H. A., Oltman, P. K., Raskin, E., & Karp, S. A. (1971). *A manual for the group embedded figures test.* Palo Alto, CA: Consulting Psychologist Press.

APPENDIX A

Resources for Educators, Parents, and Students

American Society for Engineering Education. (n.d.) *eGFI: Dream up the future.* Retrieved from http://teachers.egfi-k12.org

Exploratorium. (2016). *Education.* Retrieved from http://www.exploratorium.edu/education

James Dyson Foundation. (2016). Retrieved from http://www.jamesdysonfoundation.com

LEGO. (2016). *LEGO education.* Retrieved from https://education.lego.com/en-us

Museum of Science, Boston. (2016). *Engineering is elementary.* Retrieved from http://www.eie.org

Nast, P. (2002–2015). The 10 best STEM resources: Science, technology, engineering & mathematics resources for prek-12. *National Education Association.* Retrieved from http://www.nea.org/tools/lessons/stem-resources.html

National Science Foundation. (n.d.). *Engineering classroom resources.* Retrieved from https://www.nsf.gov/news/classroom/engineering.jsp

National Science Teachers Association. (2014). Classroom resources. *NGSS@ NSTA.* Retrieved from http://ngss.nsta.org/Classroom-Resources.aspx

PBS LearningMedia. (2016). *Teaching NGSS engineering design through media.* Retrieved from http://www.pbslearningmedia.org/collection/ngss-eng

PBS Kids. (n.d.). Resources. *Design Squad Global.* Retrieved from http://pbskids.org/designsquad/parentseducators/workshop/resources.html

Project Lead the Way. (2014). Retrieved from https://www.pltw.org

Teach Engineering: Curriculum for K–12 teachers. (n.d.) Retrieved from https://www.teachengineering.org

TryEngineering. (2016). *Lesson plans.* Retrieved from http://tryengineering.org/lesson-plans

APPENDIX B

Engineering in Formal and Informal Environments

ENGINEERING IS ELEMENTARY: CONSIDERATIONS WHEN IMPLEMENTING FOR HIGH-ABILITY LEARNERS

Ann Robinson

STEM Starters+ at the University of Arkansas at Little Rock was initiated through U. S. Department of Education funding as a Jacob K. Javits demonstration project. STEM Starters+ serves all students in grades 1 or 2 and identified gifted and talented students in grades 2 through 5 depending on the building and program configuration in school districts.

Designed to be implemented in the formal setting of a school, STEM Starters+ scales up a previous project that resulted in achievement gains in identified gifted students (Robinson, Dailey, Hughes, & Cotabish, 2014), the general cohort of elementary students (Cotabish, Dailey, Robinson, & Hughes, 2013) and elementary teachers (Dailey & Robinson, 2016). STEM Starters+ scaled up by adding another grade level (grade 1) and by infusing engineering into its STEM curriculum.

The program includes three evidence-based curricular components: William and Mary science units, Blueprints for Biography: STEM Series developed at the University of Arkansas at Little Rock, and Engineering is Elementary (EiE) units from the Museum of Science, Boston. The William and Mary science units are structured around a problem-based learning model. The Blueprints for Biography are teaching guides linked to a specific trade book biography of an engineer, inventor, or scientist and provide discussion questions and enrichment activities. The EiE units focus on the engineering design process embedded in the context of various kinds of engineering. For example, the EiE unit recommended in STEM Starters + for grade 1 students is on acoustical engineering (Robinson, Adelson, & Kidd, 2016).

Through face-to-face and online delivery methods, professional development for teachers is an integral part of the STEM Starters+ program. Teachers are prepared through face-to-face summer institutes and coached both face-to-face and online throughout the academic year. Because the evidence-based curriculum is widely available and the professional development protocols are well documented in publications, STEM Starters+ is replicable in schools.

THE MARYVILLE UNIVERSITY SCIENCE AND ROBOTICS PROGRAM FOR HIGH ABILITY STUDENTS

Steve V. Coxon

The Maryville University Science and Robotics Program for High Ability Students in St. Louis, MO, offers academic year classes, outreach into high-poverty districts, and approximately 80 classes each summer. The program serves students in preschool through high school in science, technology, engineering, art, and math (STEAM), with a focus on robotics. The program was designed for high-ability students. Held in a university setting with highly differentiated class offerings that are 1–3 or more years in advance of traditional age expectations, the program features an overall faculty/staff to student ratio lower than 1 to 6. Highly qualified faculty, including university professors, technology industry professionals, and teachers with advanced degrees in gifted education and related fields, are coupled with these small class sizes and cutting edge technology. Much like college, students can choose up to four classes to suit their interests.

Along with the summer program, which served approximately 600 students in July 2016, the Maryville University Science and Robotics Program also offers weekend classes to extend learning into the academic year. The program also offers significant outreach into high-poverty districts in the St. Louis region, including Ferguson. The program outreach has helped start Makerspaces in school libraries, implemented programming in community centers, and supported Junior FIRST LEGO League teams and FIRST LEGO League teams.

The Maryville University Science and Robotics Program serves as a research vehicle, a platform for teacher professional development, a practicum site for graduate students seeking state certification in gifted education, and a platform for preservice teacher STEM training. Recent studies on STEM talent development in precollegiate students have included measuring changes in spatial ability in robotics classes, measuring changes in creativity in robotics classes, assessing correlations between media use and creativity, and discovering parental gender stereotyping and STEM education decisions. Several teachers volunteer to serve in classrooms each summer and receive simultaneous STEM professional development. Graduate students seeking their state certification in gifted education are able to complete their practica in the program, often field-testing their own STEM curriculum units with gifted students. Each year, approximately 20 undergraduate education majors work as assistant teachers in the program, learning a variety of STEM materials and curricula while working with expert teachers. You can find out more about the Maryville University Science and Robotics Program at https://www.maryville.edu/robot.

PROJECT M²: MENTORING YOUNG MATHEMATICIANS
Kathy Gavin and Linda Sheffield

Project M²: Mentoring Young Mathematicians (Gavin, Casa, Chapin, & Sheffield, 2003–2017) is a series of curriculum units for students in grades K–2 initially developed under a National Science Foundation research grant. Similar to Project M³ units, there is a strong focus on in-depth study of advanced math concepts, mathematical creativity, and challenging and motivating students. The content across grade levels includes, geometry and measurement units (available from Kendall Hunt) and number sense and algebraic thinking units (in development and available in 2017). With a focus on high-end learning and exemplary gifted education practices, the units were field-tested with all grade-level students and are differentiated accordingly. During a national field test, research results showed Project M^2 students at each grade level made significant gains on the Iowa Tests of Basic Skills and open-ended assessments and significantly outscored comparison groups (Gavin, Casa, Adelson, & Firmender, 2013; Gavin, Casa, Firmender, & Carroll, 2013). Furthermore, Firmender (2011) found that high-ability students had significantly higher scores than their counterparts in the comparison group. These units have won the Distinguished Curriculum Award from the National Association for Gifted Children for 3 consecutive years and are widely used throughout the United States. For further information, including descriptions of the contents of each unit, visit http://www.projectm2.org.

PROJECT M^3: MENTORING MATHEMATICAL MINDS

Kathy Gavin and Linda Sheffield

Project M^3: Mentoring Mathematical Minds (Gavin, Chapin, Dailey, & Sheffield, 2003–2015) is a series of 15 curriculum units for gifted students in grades 3–6 that was initially developed under a U.S. Department of Education Javits Grant. These materials foster in-depth understanding of advanced mathematical concepts by challenging and motivating students to discuss and solve high-level problems in a fashion similar to practicing mathematicians. The curriculum design follows the tenets of *The Multiple Menu Model: A Practical Guide for Developing Differentiated Curriculum* and *The Parallel Curriculum, A Design to Develop High Potential and Challenge High-Ability Learners*. Research results have shown statistically significant gains on the Iowa Tests of Basic Skills and open-ended questions taken from released items on the Trends in International Mathematics and Science Study (TIMSS) and the National Assessment of Educational Progress (NAEP). Also Project M^3 students significantly outperformed a comparison group of like-ability peers from the same schools on these measures. These units have won the Distinguished Curriculum Award from the National Association for Gifted Children for 6 consecutive years. The project team was awarded the Research Paper of the Year for their 2009 article reporting the research results of the national field test in *Gifted Child Quarterly*. Published by Kendall Hunt, the units are currently being used to meet the needs of talented elementary students in all 50 states and in several other countries including

Singapore and Hong Kong. For further information including description of the contents of each unit, visit http://www.projectm3.org.

OPEN WINDOW SCHOOL
Adrienne Gifford, Director of Innovation & Technology

Open Window School is an independent school for gifted learners in grades K–8 in Bellevue, WA. Designed to promote critical thinking and problem solving skills, the Open Window School curriculum has a primary goal of fulfilling gifted students' deep need for student-generated inquiry and connection to sophisticated real-world problems and solutions. Students regularly engage in scientific discourse with their peers, as well as with professional scientists, technologists, and researchers in a variety of fields.

A few recent examples of student-driven innovation at Open Window School:

- Through the Student Spaceflight Experiments Program, students worked in teams to develop research proposals for microgravity experiments. Each team collaborated with professional researchers and scientists as they developed their proposal. One student-designed experiment, "Arabidopsis Germination in Martian Soil Simulant," has been selected to be conducted by astronauts aboard the International Space Station. Students hope the findings of their experiment will contribute to the development of food sources for a future human colony on Mars.

- Students worked with mentors from Near Space Systems to design their own near space experiments, assemble and program flight computers and weather sensors, and construct satellites. After launching their satellites to more than 93,000 feet, students completed advanced analysis of the collected data and shared their findings with the community at our school STEM Day.

- As part of the Verizon Innovative App Challenge, students developed mobile apps to address needs in our local community. The student-designed app HikeAbout was selected as "Best in Nation," and our students worked with a MIT-provided mentor to develop the app and make it available in the Google Play store. The students hope their app will contribute to the health and safety of hikers in our local community.

Open Window School uses the Next Generation Science Standards for our STEM curriculum benchmarks and as the paramount performance expectations in our curriculum design process. The school recognizes the importance of real-world, authentic learning experiences for the engagement and achievement of gifted students, and incorporates these experiences throughout the K–8 curriculum. You can find out more about Open Window School at http://www.openwindowschool.org.

STEMULATE ENGINEERING ACADEMY

Debbie Dailey and Alicia Cotabish

STEMulate Engineering Academy is a summer camp at the University of Central Arkansas for local elementary students. The academy serves as a practicum experience for candidates seeking a gifted and talented licensure. In collaboration with engineers and STEM experts, candidates develop and pilot curriculum integrated with engineering practices. The candidates use a constructivist, inquiry-based approach that gives participants the chance to build their own understanding while engaging in activities aligned with the state's math and science standards. Students act as real-world engineers as they dig into problems and design solutions. Additionally, engineers from Kimberly Clark Corporation collaborate with teachers and provide a real-world experience for the students. To provide opportunities to typically underrepresented groups, Kimberly Clark Corporation also supplements the cost of attendance. As a service, the academy offers free professional development to area teachers. Teachers attend the preliminary training workshop and observe the camp in progress.

REFERENCES

Cotabish, A., Dailey, D., Robinson, A., & Hughes, G. (2013). The effects of a STEM intervention on elementary students' science knowledge and skills. *School Science and Mathematics, 113,* 215–226.

Dailey, D., & Robinson, A. (2016), Elementary teachers: Concerns about implementing a science program. *School Science and Mathematics, 116,* 139–147. doi:10.1111/ssm.12162

Firmender, J. M. (2011). A study of teachers' pedagogical content knowledge and instructional practices during and after implementation of advanced primary mathematics curriculum (Doctoral dissertation). Retrieved from DigitalCommons@UConn. (AAI3475520)

Gavin, M. K., Casa, T. M., Adelson, J. L., & Firmender, J. M. (2013). The impact of challenging geometry and measurement units on the achievement of grade 2 students. *Journal for Research in Mathematics Education, 44,* 478–509.

Gavin, M. K, Casa, T. M., Firmender, J. M., & Carroll, S. R. (2013). The impact of advanced geometry and measurement curriculum units on the mathematics achievement of first-grade students. *Gifted Child Quarterly, 57,* 71–84.

Gavin, M. K., Casa, T. M., Chapin, S., & Sheffield, L. (2003–2017). *Mentoring young mathematicians: Project M²: Grades K–2.* Dubuque, IA: Kendall Hunt.

Gavin, M. K., Chapin, S., Dailey, J., & Sheffield, L. (2003–2015). *Mentoring mathematical minds: Project M³: Grades 3–6.* Dubuque, IA: Kendall Hunt.

Robinson, A., Adelson, J. L., & Kidd, K. A. (2016). *A talent for tinkering: Developing talents in young low-income children through engineering curriculum.* Manuscript in preparation.

Robinson, A., Dailey, D., Hughes, G., & Cotabish, A. (2014). The effects of a science-focused STEM intervention on gifted elementary students' science knowledge and skills. *Journal of Advanced Academics, 25,* 189–213. doi:10/1177/1932202XI4533799

About the Editors

Alicia Cotabish, Ed.D., is an associate professor in the Department of Teaching and Learning at the University of Central Arkansas. She is president-elect of The Association of the Gifted (a division of the Council for Exceptional Children) and the immediate past-president of the Arkansas Association of Gifted Education Administrators. She has authored, coauthored, and contributed to five books and a number of journal articles, book chapters, and products focused on K–20 STEM and gifted education.

Debbie Dailey, Ed.D., is an assistant professor in the Department of Teaching and Learning at the University of Central Arkansas, where she coordinates the gifted and talented education graduate program. Debbie has authored multiple publications and presented numerous presentations and workshops focused on K–12 STEM and gifted education. Additionally, she directs a new summer camp at her university for elementary students, STEMulate Engineering Academy. Before working in higher education, Debbie taught for 20 years in the public school system, including 14 years as a high school science teacher and 6 years as a gifted and talented elementary and middle school teacher.

About the Authors

Cheryll M. Adams, Ph.D., is the Director Emerita of the Center for Gifted Studies and Talent Development at Ball State University. She teaches online courses in gifted education for the University of Virginia and has presented widely at local, state, national, and international conferences. She is a former member of the Board of Directors of NAGC and is currently the chair of the NAGC Professional Standards Committee. She has many publications, including 12 coauthored books, 16 book chapters, and others.

Jill L. Adelson, Ph.D., is an associate professor in the Educational Psychology, Measurement, and Evaluation program at the University of Louisville. She earned a joint Ph.D. in gifted education and in measurement, evaluation, and statistics from the University of Connecticut, and she previously taught self-contained gifted and talented fourth grade. She is an associate editor for the *Journal of Advanced Academics*. Dr. Adelson is currently working on three Javits projects—Project SPARK, STEM Starters+, and Reaching Academic Potential. Her research focuses on applications of advanced methods in gifted education research.

Kinnari Atit, Ph.D., is a postdoctoral researcher in the Department of Psychology at Northwestern University. Her areas of research include the intersection of spatial thinking and science, technology, engineering, and math (STEM) education, and the identification of academically talented students. Her current work focuses on understanding the role of spatial thinking skills in STEM domains, and also how to develop those skills in students both in and out of the classroom. Prior to Northwestern, Kinnari was a postdoctoral researcher/program coordinator at the Johns Hopkins University Center for Talented Youth where she investigated alternative methods of identifying academic talent in historically underrepresented students.

Gina Howes Boshears is an Advanced Practice Partner in the Trauma Unit at the University of Arkansas for Medical Sciences. After pursuing a bachelor's degree in biological sciences, Gina graduated with a BSN from California State University, Sacramento, and obtained a master's degree in science communication at Emerson College in Boston, MA. She has worked with multiple organizations throughout the Boston area, and since her move to Hot Springs, AR, she has served for 6 years on the Mid-America Science Museum Board of Directors while working toward a Ph.D. in science education through the University of Arkansas at Fayetteville.

Michelle B. Buchanan is the science education program coordinator for the University of Central Arkansas STEMTeach Program. She has been teaching for 18 years and taught for 14 years in the public school system as a junior high school science and engineering teacher. She is the recipient of several national awards for her teaching, curriculum, and lessons. Michelle is a National Board Certified Teacher in early adolescence sciences and a K–12 gifted and talented certified teacher. She also teaches project-based instruction, classroom diversity and differentiation, and educational psychology and pedagogy for the STEMTeach program.

Scott A. Chamberlin, Ph.D., is a professor at the University of Wyoming in the field of mathematics education and the department head of Elementary and Early Childhood Education. His research interests include the use of problem-solving activities with upper elementary and middle grade gifted students. Specifically, he investigates student affect as it relates to mathematical problem solving. He created the Chamberlin Affective Instrument for Mathematical Problem Solving (CAIMPS), which may be used by classroom teachers to formally assess student affect as they complete problem-solving tasks in mathematics.

Steve V. Coxon, Ph.D., is an associate professor and Director of Gifted Education Programs at Maryville University, including the gifted education graduate program; the Maryville Young Scholars Program to increase gifted program diversity; the Children using Robotics for Engineering, Science, Technology, and Math (CREST-M) math-focused STEM curricula project; and the Maryville Science and Robotics Program. He also directs the STEM Education Certificate Program. Steve conducts research on developing STEM talents and is author of numerous publications including the book *Serving Visual-Spatial Learners*. He serves as the science education columnist for *Teaching for High Potential* and book review editor for *Roeper Review*.

Laurie J. Croft, Ph.D., is a clinical associate professor in the Department of Teaching and Learning at the University of Iowa College of Education and is the associate director for professional development at The Connie Belin & Jacqueline

N. Blank International Center in Gifted Education and Talent Development, a part of the College of Education. Laurie has made presentations at various state, national, and international conferences, and to parent groups, teachers, and school boards. She also has experience facilitating professional learning in gifted education for educators from around the world.

April DeGennaro, Ph.D., has taught gifted students for 27 years and for the past 15 has taught at Peeples Elementary in Peachtree City, GA. She currently facilitates a "SCREAM lab" at Peeples Elementary, where science, coding, research, engineering, animation, and math provide authentic, real-world challenges and differentiation for gifted students. In 2015, April was awarded the Georgia Gifted Teacher of the Year by the Georgia Association for Gifted Children. She is the incoming chair of the Global Awareness Network for NAGC and part of the leadership team for ISTE's Learning Spaces Network. April is a Code.org facilitator for CS Fundamentals and CS in Science and provides computer science professional learning to K–8 Georgia educators.

Umadevi Garimella, Ph.D., is the director of the University of Central Arkansas STEM Institute and a College Board Consultant for AP Biology. She earned her Ph.D. in botany and double master's degrees in botany and chemistry. Umadevi has 14 years of successful college-level teaching and research. She published 23 scientific research papers in peer-reviewed journals and given several presentations at professional meetings. As a director of the STEM Institute, Umadevi is responsible for the development, use and promotion of research-proven pedagogies, and rigorous content in P–16 STEM education in Arkansas.

Brian Housand, Ph.D., is an associate professor and the co-coordinator of the Academically and Intellectually Gifted Program at East Carolina University. Brian earned a Ph.D. in educational psychology at the University of Connecticut's Neag Center for Gifted Education and Talent Development with an emphasis in both gifted education and instructional technology. He serves on the National Association for Gifted Children's Board of Directors as a Member-At-Large. He researches ways in which technology can enhance the learning environment and is striving to define creative productive giftedness in a digital age.

Kristy Kidd is the director of STEM Starters+, a Jacob K. Javits Project, at the University of Arkansas at Little Rock. Kristy has 21 years of experience teaching grades K–8 science in Little Rock Public Schools. She has served as building gifted and talented facilitator for Little Rock Schools, an adjunct professor of early childhood science methods for the University of Arkansas at Little Rock, and K–12 math and science specialist for eStem Public Charter Schools. Kristy is the recipient of the Milken National Educator Award, a state finalist for the Presidential Award for Excellence in Mathematics and Science Teaching, and the recipient of the Arkansas Museum of Discovery SPARK! Star Award, which honors Arkansans who have been influential in STEM fields.

Irene Lee is a research scientist in MIT's Scheller Teacher Education Program and Education Arcade. She is the founder and director of Project GUTS: Growing Up Thinking Scientifically and was the principal investigator of Project GUTS, New Mexico Computer Science for All, GUTS y Girls and Yo-GUTC. Irene serves as the chair of the Computer Science Teachers Association (CSTA) Computational Thinking Task Force and is a member of the CSTA K–12 Computer Science Standards writing team and the K12 CS Framework writing team.

Bronwyn MacFarlane, Ph.D., is a professor of gifted education at the University of Arkansas at Little Rock. She served as associate dean for the College of Education and Health Professions, as chair-elect of the STEM Network, and past-chair of the Counseling Network for the National Association for Gifted Children. Her edited book, *STEM Education for High-Ability Learners: Designing and Implementing Programming* (2016), is the first of its kind to bring together discussion of the critical elements for delivering STEM programming to develop high-ability talent in the STEM fields.

Eric L. Mann, Ph.D., is an assistant professor of mathematics education at Hope College in Holland, MI. After completing a military career in the Air Force, he taught elementary and middle school for 7 years before entering the doctoral program in educational psychology at the University of Connecticut with emphasis in gifted and mathematics education. Eric served on the faculty at Purdue University's Institute for P–12 Engineering Research and Learning and the Gifted Education Resource Institute before accepting his current position. He is interested in a deeper understanding of creativity and talent development within the STEM disciplines.

Rebecca L. Mann, Ph.D., joined the faculty at Hope College in 2013 after serving as clinical assistant and associate professor at Purdue University for 8 years. She also teaches online gifted education courses for the University of Connecticut, serving as the professor for Integrating STEM Disciplines for Gifted & Talented Students. Prior to her doctoral studies, she worked for 17 years as an elementary and middle school classroom teacher, as well as a gifted resource teacher, in Colorado and New Hampshire. In 2001, she was named the New Hampshire Educator of the Year of the Gifted.

Rachelle Miller, Ph.D., is an assistant professor in the Department of Teaching and Learning at the University of Central Arkansas. She works in the Gifted and Talented Education program teaching Affective Strategies for the Gifted and Talented. She collaborates with Arkansas A+ Schools, completing program evaluation and assisting in the development of arts integrated curriculum in participating schools. Her research interests include supporting the academic needs of low-income gifted students, integrating the arts into STEM and gifted curriculum, and examining teacher perceptions of arts integration.

Nielsen Pereira, Ph.D., is an assistant professor of Gifted, Creative, and Talented Studies at Purdue University. His research interests include the design and assessment of learning in varied gifted and talented education contexts, understanding gifted and talented student experiences in talent development programs in and out of school, and conceptual, contextual, and measurement issues in the identification of gifted and talented populations. He codeveloped, with engineering education colleagues, the curriculum for the STEAM Labs program, which challenges middle and high school students to learn and apply the engineering design process in a cooperative learning environment. He is a regular presenter at national and international conferences on educational research, gifted education, and STEM education. He currently serves as Associate Editor for *Gifted and Talented International* and is past editor of *Mosaic*. He taught English as a second language for 12 years in public schools and language institutes in Brazil and was coordinator of student programs in the Gifted Education Resource Institute at Purdue University.

Paula M. Olszewski-Kubilius, Ph.D., is the Director of the Center for Talent Development at Northwestern University and a professor in the School of Education and Social Policy. Over the past 30 years, she has created programs for all kinds of gifted learners and written extensively about talent development. She has served as the editor of *Gifted Child Quarterly*, as coeditor of the *Journal of Secondary Gifted Education*, and on the editorial boards of *Gifted and Talented International*, *Roeper Review*, and *Gifted Child Today*. She is past-president of the National Association for Gifted Children and received the Distinguished Scholar Award in 2009 from NAGC.

Kay E. Ramey is a Ph.D. candidate in Learning Sciences at Northwestern University. Her current research interests include the role of external representations, technology, and spatial thinking in K–12 engineering and integrated STEAM (science, technology, engineering, arts, and math) learning.

Ann Robinson, Ph.D., is professor and founding director of the Jodie Mahony Center at the University of Arkansas at Little Rock, where she coordinates graduate programs in gifted education. She is a former editor of *Gifted Child Quarterly*, is a past-president of the National Association for Gifted Children, and received the Early Leader, the Early Scholar, the Distinguished Service, and the Distinguished Scholar awards from NAGC. Ann has generated more than $25 million dollars in external funding, including five Jacob K. Javits projects. A recent Javits project, STEM Starters, was identified by the National Science Teachers Association as exemplary.

Callie Slider received her Masters of Art in Teaching at the University of Central Arkansas and Masters of Art in Art History at the University of Arkansas, Little Rock. She teaches elementary art in the Little Rock School District, emphasizing the new National Arts Education Standards through choice-based art. She collaborated with the Gifted and Talented Education program at the University of Central Arkansas and Arkansas A+ Schools to create an arts integrated curriculum. She acts as a mentor to promote choice-based art, also known as Teaching for Artistic Behaviors, by providing professional development for teachers within her district and state.

Jason Trumble, Ph.D., is an assistant professor of education at the University of Central Arkansas, where he teaches preservice teachers to effectively integrate technology, content, and pedagogy. His research is centered around the intersection of teaching and technology with a particular focus on emerging technologies for education. Prior to his current appointment, he taught in K–12 classrooms in California and Texas. For more than a decade, Jason has been dedicated to differentiating curriculum and connecting exceptional teaching to content through effectively integrating technology.

David Uttal, Ph.D., is a professor of psychology and education at Northwestern University. His research focus is on cognitive development and education, particularly early symbolic thought and spatial cognition. He studies how children's spatial thinking develops and how it can be enhanced. He is a Fellow of both the American Psychological Association and the Society for Psychological Science. His meta-analysis of methods for improving spatial thinking received the 2014 George Miller Award from the Psychological Association for the best paper in general psychology.

Joyce VanTassel-Baska, Ed.D., is the Jody and Layton Smith Professor Emerita of Education and founding director of the Center for Gifted Education at William & Mary, where she developed a graduate program and a research and development center in gifted education. She also initiated and directed the Center for Talent Development at Northwestern University. Joyce has published widely including numerous books and more than 600 refereed journal articles, book chapters, and scholarly reports.

Krissy Venosdale is the Innovation Coordinator at Kinkaid Lower School in Houston, TX, where she is working to develop the Pre-K–grade 4 Makerspace, "The Launch Pad." With more than 10 years of experience in teaching gifted learners, she loves incorporating creativity and technology in authentic ways, supporting teachers, and seeing kids experience the true power of global connections.